of the National Institute of Building Sciences

Homebuilders' Guide to Earthquake-Resistant Design and Construction

FEMA 232 — June 2006

Prepared by the
Building Seismic Safety Council
for the
Federal Emergency Management Agency
of the Department of Homeland Security

National Institute of Building Sciences
Washington, D.C.

NOTICE: Any opinions, findings, conclusions, or recommendations expressed in this publication do not necessarily reflect the views of the Federal Emergency Management Agency. Additionally, neither FEMA nor any of its employees make any warranty, expressed or implied, nor assume any legal liability or responsibility for the accuracy, completeness, or usefulness of any information, product, or process included in this publication.

The opinions expressed herein regarding the requirements of the *International Residential Code* do not necessarily reflect the official opinion of the International Code Council. The building official in a jurisdiction has the authority to render interpretation of the code.

This report was prepared under Contract EMW-1998-CO-0419 between the Federal Emergency Management Agency and the National Institute of Building Sciences.

PREFACE

The Federal Emergency Management Agency (FEMA), which is part of the Department of Homeland Security, works to reduce the ever-increasing cost that disasters inflict on the nation. Preventing losses before they occur by designing and constructing buildings and their components to withstand anticipated forces from various hazards is one of the key components of mitigation and is one of the most effective ways of reducing the cost of future disasters.

The National Earthquake Hazards Reduction Program (NEHRP) is the federal program established to address the nation's earthquake threat. NEHRP seeks to resolve two basic issues: how will earthquakes affect us and how do we best apply our resources to reduce their impact on our nation. The program was established by Congress under the Earthquake Hazards Reduction Act of 1977 (Public Law 95-124) and was the result of years of examination of the earthquake hazard and possible mitigation measures. Under the NEHRP, FEMA is responsible for supporting program implementation activities, including the development, publication, and dissemination of technical design and construction guidance documents.

Generally, there has not been much technical guidance addressing residential buildings unless they are located in areas of high seismicity or exceed a certain size or height. This is because most residential buildings were thought to perform fairly well in earthquakes due to their low mass and simple construction. While buildings may not normally experience catastrophic collapse, they can still suffer significant amounts of damage, rendering them uninhabitable. This is especially true when construction techniques are less than adequate. What is particularly important from FEMA's point of view is that, given the sheer number of this type of building, even minor damage represents a significant loss potential and temporary housing demand that will need to be addressed after an earthquake by all levels of government.

After the San Fernando earthquake in 1971, a study of residential buildings and the damage they suffered was conducted by a team of experts under funding from the Department of Housing and Urban Development (HUD) and the National Science Foundation (NSF). HUD utilized these data to develop a non-engineering document entitled *Home Builder's Guide to Earthquake Design*. This manual, originally published in 1980, provided easy-to-follow information to the average homebuilder on steps for reducing potential earthquake damage. In July 1992 it was reprinted as a joint FEMA-HUD document, also known as FEMA 232. The manual was subsequently updated for FEMA and the revised publication, *Home Builders Guide to Seismic Resistant Construction* (FEMA 232), was published in August 1998.

Since that time, there have been several significant changes that needed to be incorporated into this document to keep it current. The first and most important change was the completion of the FEMA-funded Consortium of Universities for Research in Earthquake Engineering (CUREE)-Caltech Woodframe Project. This project was funded using FEMA Hazard Mitigation Grant Program funds available after the Northridge earthquake and was designed to address the unexpected amount of damage suffered by wood frame residential structures. Similar to the successful FEMA/SAC Steel Moment Frame Buildings Project, this effort combined academic research and testing of wood frame buildings and components with the development of engineering-based design guidance for future construction. The project yielded some interesting findings that needed to be captured in a guidance document.

A second change was the development and publication of the 2000 *International Residential Code* (*IRC*) by the International Code Council. This model residential building code replaced the Council of American Building Officials (CABO) *One- and Two-Family Dwelling Code*, which did not adequately address earthquake loads. The *IRC* reflects on the *NEHRP Recommended Provisions* and is intended to adequately address the earthquake hazard.

This publication presents seismic design and construction guidance for one- and two-family houses in a manner that can be utilized by homebuilders, knowledgeable homeowners, and other non-engineers. It incorporates and references the prescriptive provisions of the 2003 *International Residential Code* as well as the results of the FEMA-funded CUREE-Caltech Woodframe Project. The manual includes prescriptive building detail plans based on state-of-the-art earthquake-resistant design for use by homebuilders and others in the construction of a non-engineered residential structure. Further, the manual also uses the results of recent loss investigations as well as current research and analysis results to identify a number of specific **above-code** measures for improved earthquake performance along with their associated costs. A typical modern house is used to illustrate the application and benefits of **above-code** measures. This manual replaces the *Home Builders Guide to Seismic Resistant Construction* (FEMA 232) published by FEMA in August 1998 as well as earlier FEMA and HUD versions.

Finally, FEMA wishes to express its deepest gratitude for the significant efforts of primary authors J. Daniel Dolan, Kelly Cobeen, and James Russell; the members of the Project Team; the many reviewers and workshop attendees; and the BSSC Board of Direction and staff. Their dedication and hard work made this document possible.

Department of Homeland Security/Federal Emergency Management Agency

ACKNOWLEDGEMENTS

The Building Seismic Safety Council is grateful to all those involved in developing this updated guide to earthquake-resistant home design and construction. As chairman of the BSSC Board of Direction, it is my pleasure to express appreciation to the members of the team developing this guide:

> J. Daniel Dolan, PhD, PE, Professor, Washington State University, Wood Materials and Engineering Laboratory, Pullman (team leader)
>
> Kelly Cobeen, Structural Engineer, Cobeen and Associates Structural Engineering, Lafayette, California
>
> James E. Russell, Building Codes Consultant, Concord, California

Special thanks also go to Gerald Jones, retired building official, and Douglas Smits, the chief building official of Charleston, South Carolina, who served with me and the writers on the committee overseeing this project.

The BSSC also is grateful to those who participated in a workshop discussion (Appendix F) of an early draft of this guide and willingly shared their experiences with respect to earthquake-resistant home building. Special thanks go to those individuals who reviewed and commented on subsequent drafts of this document; their input has made this a much more useful guide.

I also wish to thank the members of the BSSC Board of Direction who have recognized the importance of this effort and provided sage advice throughout the project. Special thanks also are due to the BSSC staff who worked untiringly behind the scenes to coordinate the project resulting in this report.

Finally the BSSC is grateful to Michael Mahoney, FEMA project officer, for his insight and ongoing support, to the International Code Council for their work in reviewing and disseminating this guide, and to the National Association of Home Builders for their assistance in disseminating this guide.

Jim. W. Sealy
Chairman, BSSC Board of Direction

TABLE OF CONTENTS

Preface iii

Acknowledgements v

Executive Summary 1

Chapter 1 Introduction 3
    1.1 Background 3
    1.2 Above-code Recommendations 4
    1.3 The *International Residential Code* 5
    1.4 *IRC* Seismic Design Categories 5
    1.5 Building Response 10
    1.6 Important Concepts Concerning Homesite 12
    1.7 Model House Used for Guide Analysis 14
    1.8 Other Hazards 17

Chapter 2 Earthquake-Resistance Requirements 21
    2.1 *IRC* General Earthquake Limitations 21
    2.2 Load Path 22
    2.3 House Configuration Irregularities 29

Chapter 3 Foundations and Foundation Walls 49
    3.1 General Foundation Requirements 49
    3.2 Concrete Foundations 53
    3.3 Masonry Foundations 56
    3.4 Footing Width 57
    3.5 Special Soil Conditions 58
    3.6 Foundation Resistance to Sliding from Lateral Loads 59
    3.7 Special Considerations for Cut and Fill Sites 61
    3.8 Foundation Walls 62
    3.9 Foundation Wall Thickness, Height, and Required Reinforcing 63
    3.10 Wood-Framed Wall Bottom Plate and Foundation Sill Plate Anchorage ... 64
    3.11 Required Location for Anchor Bolts along Exterior Walls 65
    3.12 Required Anchorage along Interior Braced Walls 65
    3.13 Anchoring of Interior Braced Walls in SDCs D_1 and D_2 67

Chapter 4 Floor Construction 69
    4.1 General Floor Construction Requirements 69
    4.2 Wood-Framed Floor Systems 70
    4.3 Cantilevered Floors 71
    4.4 Requirements for Blocking 73
    4.5 Connection of Floor Joists to Wall Top Plate or Foundation Sill Plate Below 75
    4.6 Floor Sheathing 77

  4.7 Lateral Capacity Issues for Wood Framed Floors Using Wood Structural Panels............77
  4.8 Concrete Slab-on-Grade Floors 81

Chapter 5 Walls..**83**
  5.1 Wood Light-Frame Construction..83
  5.2 Stone and Masonry Veneer..103
  5.3 Cold-formed Steel Houses 107
  5.4 Masonry Wall Houses 110
  5.5 Insulating Concrete Form (ICF) Wall Houses..115

Chapter 6 Roof-Ceiling Systems..**121**
  6.1 General Roof-Ceiling Requirements 121
  6.2 Special Framing Considerations...122
  6.3 Blocking and Lateral Load Paths for Roof Systems...124
  6.4 Connection of Ceiling Joists and Rafters to Walls Below..126
  6.5 Roof Sheathing 126
  6.6 Lateral Capacity Issues for Wood Framed Roofs...128
  6.7 Quality Control 129

Chapter 7 Chimneys, Fireplaces, Balconies, and Decks..**131**
  7.1 Chimneys and Fireplaces..131
  7.2 Balconies and Decks...137

Chapter 8 Anchorage of Home Contents..**141**
  8.1 General..141
  8.2 Water Heater Anchorage 141
  8.3 Securing Other Items 144

Chapter 9 Existing Houses **151**
  9.1 Additions and Alterations 151
  9.2 Earthquake Upgrade Measures...154

APPENDICES

Appendix A Analysis of Model House Used in This Guide..**167**
  A1 Model House...167
  A2 Analysis Using Standard Engineering Design Methods...171
  A3 Analysis Using Nonlinear Methods..171
  A4 Analysis Results..174

Appendix B Earthquake Provisions Checklist for Builders and Designers **179**

Appendix C Earthquake Provisions Checklist for Designers and Plan Checkers................**187**

Appendix D Significant Changes for the 2006 *International Residential Code* **193**

D1 Revised Seismic Design Maps..193
D2 Addition of Seismic Design Category D_0...193
D3 Change in Applicability of Irregular Building Requirements 199
D4 Clarification and Addition of Requirement for Masonry Veneer..................................199

Appendix E References and Additional Resources..201

Appendix F Homebuilders' Guide Project Participants...207

LIST OF FIGURES

Figure 1-1 Seismic Design Categories.. 6-9
Figure 1-2 Forces induced in a house due to earthquake ground motion 10
Figure 1-3 Concept of actual vs. code seismic forces...11
Figure 1-4 Concept of ductility...11
Figure 1-5 Definition of drift..12
Figure 1-6 A house damaged due to fault movement...13
Figure 1-7 Houses damaged due to a landslide at the site 13
Figure 1-8 Elevation of model house used as an example for this guide 14
Figure 1-9 Model house floor plan 15

Figure 2-1 Chain illustrating load path concept..22
Figure 2-2 Lateral loads induced in a building due to wind or earthquakes.......................23
Figure 2-3 Loading and deflection of roof, ceiling, and floor systems...............................24
Figure 2-4 Loading and deflection of bracing wall systems...25
Figure 2-5 Load transfer between components in a building 26
Figure 2-6 Horizontal load path connections and deformations...27
Figure 2-7 Overturning load path connections and deformations 28
Figure 2-8 Depiction of ideal building configuration for earthquake resistance 30
Figure 2-9 Irregular building configuration with open front 31
Figure 2-10 Rotational response and resistance to torsion 31
Figure 2-11 Concentration of loads and possible damage and failure due to irregular plan 32
Figure 2-12 Roof-level damage to building with a T-shaped plan 33
Figure 2-13 Deformation pattern due to soft-story behavior 34
Figure 2-14 Change in deformation patterns associated with soft-story irregularity 34
Figure 2-15 House experiencing soft- and weak-story behavior 35
Figure 2-16 Apartment building experiencing soft- and weak-story behavior...................35
Figure 2-17 Detailing required for stepped foundation walls...36

Figure 3-1 Perimeter foundation with separately placed footing and stem wall 49
Figure 3-2 Sliding action resisted by foundation..50
Figure 3-3 Overturning action resisted by foundation..51
Figure 3-4 Interior braced wall and floor framing..52
Figure 3-5 Foundation requirements for interior braced wall line on slab-on-grade construction 52
Figure 3-6 Recommended minimum reinforcement for concrete footings and stem walls...........54

Figure 3-7 Above-code horizontal reinforcing lap at corners and intersections 54
Figure 3-8 Above-code use of vertical dowels to connect a slab-on-grade to a separately poured footing 55
Figure 3-9 Inverted-T footing dimensions 58
Figure 3-10 Lateral resistance provided by foundation 59
Figure 3-11 Above-code stepped foundation reinforcing detail 60
Figure 3-12 Foundation supported on rock and fill 61
Figure 3-13 Example of damage caused in building on cut and fill site 62
Figure 3-14 Interior braced wall on slab-on-grade 66

Figure 4-1 An unblocked floor diaphragm 69
Figure 4-2 Slab-on-grade and perimeter footing transfer loads into soil 70
Figure 4-3 Cantilevered floor restrictions 71
Figure 4-4 Cantilevered joist at braced wall above 73
Figure 4-5 Interior bearing line 74
Figure 4-6 Blocking below interior braced wall 74
Figure 4-7 Blocking floor joists spaced apart for piping in floor 75
Figure 4-8 Toe nail configuration requirements 76
Figure 4-9 Blocked diaphragm configuration 79
Figure 4-10 Diaphragm loads on long and short sides 79
Figure 4-11 Reinforcing straps at large diaphragm openings 80

Figure 5-1 Sliding, overturning, and racking action resisted by walls and foundation 84
Figure 5-2 Exploded view illustrating load paths 85
Figure 5-3 Wall action for resisting lateral loads 86
Figure 5-4 Exploded view of typical residential wall segment 86
Figure 5-5 Detailing differences for three options when using wood structural panel sheathed walls 87
Figure 5-6a Plan view of crawl space for model house in SDC C 92
Figure 5-6b Plan view of first floor for model house in SDC C 93
Figure 5-6c Plan view of second floor for model house in SDC C 94
Figure 5-7a Plan view of crawl space for model house in SDC D_2 95
Figure 5-7b Plan view of first floor for model house in SDC D_2 96
Figure 5-7c Plan view of second floor for model house in SDC D_2 97
Figure 5-8 Properly and overdriven nails 98
Figure 5-9 Illustration of concept of equal numbers of fasteners in line for symmetric nailing schedule 100
Figure 5-10 Sheathing detail for extending the sheathing over the band joist 102
Figure 5-11 Applications of masonry and stone veneer 103
Figure 5-12 Stone veneer damage during the earthquake in Northridge, California 104
Figure 5-13 Cold-formed steel house under construction 107
Figure 5-14 House constructed with masonry walls 111
Figure 5-15 Symmetric layout of walls to distribute loads uniformly and thereby prevent torsion 114
Figure 5-16 Insulated concrete form house under construction 116
Figure 5-17 Illustration of cross-grain bending of wood ledger 118

Figure 6-1 Typical light-frame roof-ceiling system 121
Figure 6-2 Ridge board for 12:12 pitch roof 123
Figure 6-3 Gable end wall or gable truss bracing 124
Figure 6-4 Blocking at rafters to exterior wall 125
Figure 6-5 Typical roof diaphragm sheathing nailing when wood structural panels are used 127

Figure 7-1 Chimney damage 131
Figure 7-2 Chimney damage in Northridge earthquake 132
Figure 7-3 Locations for earthquake anchorage of masonry chimney at exterior house wall 133
Figure 7-4 Chimney section showing earthquake anchorage 134
Figure 7-5 Anchorage detail for framing parallel to exterior wall 134
Figure 7-6 Anchorage detail for framing perpendicular to exterior wall 135
Figure 7-7 Factory-built flue and light-frame enclosure 136
Figure 7-8 Collapsed factory-built chimney, light-frame enclosure, and deck 136
Figure 7-9 Decks and balconies in residential contruction 138
Figure 7-10 Hold-down device providing positive connection of deck framing to house framing 139

Figures 8-1a & 8-1b Securing a water heater with wall bracing 142
Figures 8-2a & 8-2b Securing a water heater with corner bracing 143
Figures 8-3 a-f Reproductions of pages from FEMA 74 on securing household goods 145-150

Figure 9-1 Alterations to existing house 152
Figure 9-2 Horizontal addition and vertical addition 153
Figure 9-3 Cape Mendocino earthquake damage to cripple wall house with slab-on-grade additions 154
Figure 9-4 Common anchorage configurations 156
Figure 9-5a House with collapsed cripple walls 157
Figure 9-5b A house with severe damage due to cripple wall collapse 157
Figure 9-6 Cripple wall bracing 158
Figure 9-7 Weak and soft story bracing 159
Figure 9-8 Common open-front occurrences in one- and two-family detached houses 160
Figure 9-9 Detailing of narrow bracing wall piers at open fronts 161
Figure 9-10 House located on a hillside site damaged during Northridge, California, earthquake 162
Figure 9-11 Anchorage of floor framing to the uphill foundation 163
Figure 9-12 Split-level ties for floor framing 164
Figure 9-13 Anchorage of concrete or masonry walls to floor, roof, or ceiling framing 165

Figure A1 Slab-on-grade base 168
Figure A2 Level cripple wall base 168
Figure A3 Hillside base 168
Figure A4 Basement 168
Figure A5 Analysis model 172
Figure A6 Hyseteric parameters for model 172

Figure D1 Seismic Design Categories, Site Class D 194-198

LIST OF TABLES

Table 2-1 Load Path Connections for Horizontal Sliding 38-40
Table 2-2 Load Path Connections for Overturning... 41-43
Table 2-3 *IRC* House Configuration Irregularities 44-47

Table 3-1 Summary of 2003 *IRC* Continuous Foundation and Anchor
Bolt Requirements for Braced Wall Lines in One- and Two-family Houses 68
Table 3-2 Summary of 2003 *IRC* Continuous Foundation and Anchor
Bolt Requirements for Bearing Walls in One- and Two-family Houses 68

Table 5-1 Braced Wall Panel Construction Methods (Materials) Recognized by the *IRC* 88
Table 5-2 *IRC* Sheathing Requirements for Seismic Design Categories C, D_1, and D_2 90

Table A-1 Example Wall Bracing per 2003 *IRC*, Slab-on-grade Base Condition......................170
Table A-2 Hysteretic Parameters Used for Nonlinear Dynamic Model 173
Table A-3 Selected Results for *IRC* Bracing Provisions, Slab-on-grade Base Condition..........175
Table A-4 Selected Results for Above-Code Measures, Slab-on-grade Base Condition 176

EXECUTIVE SUMMARY

This guide provides information on current best practices for earthquake-resistant house design and construction for use by builders, designers, code enforcement personnel, and potential homeowners. It incorporates lessons learned from the 1989 Loma Prieta and 1994 Northridge earthquakes as well as knowledge gained from the FEMA-funded CUREE-Caltech Woodframe Project. It also introduces and explains the effects of earthquake loads on one- and two-family detached houses and identifies the requirements of the 2003 *International Residential Code (IRC)* intended to resist these loads. The stated purpose of the *IRC* is to provide:

> ... minimum requirements to safeguard the public safety, health, and general welfare, through affordability, structural strength, means of egress facilities, stability, sanitation, light and ventilation, energy conservation and safety to life and property from fire and other hazards attributed to the built environment.

Because the building code requirements are minimums, a house and its contents still may be damaged in an earthquake even if it was designed and built to comply with the code. Research has shown, however, that earthquake damage to a house can be reduced for a relatively small increase in construction cost. This guide identifies **above-code** techniques for improving earthquake performance and presents an estimate of their cost. Note that the information presented in this guide is not intended to replace the *IRC* or any applicable state or local building code, and the reader is urged to consult with the local building department before applying any of the guidance presented in this document.

The information presented in this guide applies only to one- and two-family detached houses constructed using the nonengineered prescriptive construction provisions of the *IRC*. Applicable *IRC* limits on building configuration and construction are described.

A typical model house is used to illustrate the concepts discussed and to identify approximate deflections under earthquake loading, which permits performance to be compared for various building configurations using the minimum code requirements and the **above-code** techniques. The **above-code** recommendations are based on an analysis of the model house as well as comparative tests performed by various researchers and the lessons learned from investigation of residential building performance in past earthquakes. A nonlinear time-history analysis was performed for the model building using the SAWS computer program developed as part of the CUREE-Caltech Woodframe Project (Folz and Filiatrault, 2002). Details of the analysis are presented in Appendix A.

Additional appendices feature checklists for builders, designers, and plan checkers; explain significant differences between the 2003 and 2006 editions of the *IRC*; and present a list of reference materials.

Chapter 1
INTRODUCTION

1.1 BACKGROUND

After the 1971 earthquake in San Fernando, California, a study of residential buildings and the types of damage they suffered was conducted by a team of experts with funding from the Department of Housing and Urban Development (HUD) and the National Science Foundation (NSF). Subsequently, HUD utilized the results of that study to develop a nonengineering guidance document entitled *Home Builder's Guide to Earthquake Design*. This manual was originally published in June 1980 to provide homebuilders with easy-to-follow guidance for reducing potential earthquake damage.

In July 1992, the Federal Emergency Management Agency (FEMA) reissued the HUD manual with some updated material as a joint FEMA/HUD publication identified as FEMA 232. By the mid-1990s, it was apparent that this publication was in need of updating, especially to take into account some of the early findings from the 1994 Northridge earthquake. This update was prepared for FEMA by SOHA Engineers and was published by FEMA in August 1998 as the *Home Builders Guide to Seismic Resistant Construction*, again as FEMA 232.

Since that time, significant events have occurred warranting another updating of the guide. First was completion of the FEMA-funded CUREE-Caltech Woodframe Project. This project, funded under the FEMA Hazard Mitigation Grant Program, addressed the unexpected damage suffered by woodframe residential structures during the Northridge earthquake. Project testing of complete woodframe buildings and individual components that resist or transmit earthquake loads yielded some interesting findings that needed to be captured in a guidance document.

Another significant event occurred in 1994 when the International Code Council was established to develop a single set of comprehensive and coordinated national model construction codes. Prior to this time, the three organizations that founded the ICC – the Building Officials and Code Administrators International, Inc. (BOCAI), the International Conference of Building Officials (ICBO), and Southern Building Code Congress International, Inc. (SBCCI) – each published a set of model building codes that generally were used in distinct regions of the nation. The initial editions of the ICC's *International Building Code* (*IBC*) and *International Residential Code* (*IRC*) were published in 2000 and updates were issued in 2003 and 2006. The *IBC* replaced BOCAI's *National Building Code*, ICBO's *Uniform Building Code*, and SBCCI's *Standard Building* Code. For prescriptive residential construction, the *IRC* replaced the *One- and Two-Family Dwelling Code* of the Council of American Building Officials (CABO). (Note that the *IBC* contains prescriptive and engineering provisions for light-frame wood construction. In certain cases, the *IBC* prescriptive provisions are different from those in the *IRC*. The National Fire Protection Association's *NFPA 5000 Building Construction and Safety Code* does not contain provisions for prescriptive residential construction.)

In order to address these and other changes, FEMA initiated a complete rewrite of the 1998 document; this guide, which retains the FEMA 232 designation, is the result. One- and two-family detached houses of light-frame wood construction are addressed; however, the discussion is relevant to other materials of construction likely to be used for houses including light-frame cold-formed steel and insulated concrete form. Explained in this guide are:

- The basic principles of earthquake-resistant design,
- The specific prescriptive seismic provisions of the 2003 *International Residential Code*,
- The results of recent research and analysis, and
- Measures exceeding code requirements that are expected to reduce the amount of damage from an earthquake (see Section 1.2 below).

The guide also includes limited guidance on applying the principles of earthquake resistance to house additions and alterations and on anchoring typical house furnishings and equipment such as hot water heaters. Appendices describe the analyses performed in developing this guide, present checklists for builders, designers, and plan checkers; explain significant differences between the 2003 and 2006 editions of the *IRC*; and identify reference materials and participants in the project resulting in this guide.

1.2 ABOVE-CODE RECOMMENDATIONS

The **above-code** recommendations included in this guide describe details that, when incorporated into a house, can be expected to result in improved performance above that expected from a house designed and constructed following the minimum requirements of the *IRC*. The **above-code** techniques reduce the deformations of the house during an earthquake and therefore reduce the amount of damage. **Above-code** recommendations are printed in boldface type in this guide and appear, with associated discussion, in boxes. While the **above-code** recommendations are expected to improve the performance of a house in an earthquake and thereby reduce damage, many will involve some added costs. The costs associated with **above-code** recommendations presented in this guide are based on an estimate prepared by a homebuilder in the Seattle area and are cited as a percentage of the basic framing cost for the model house analyzed during development of this guide. Presenting the cost increase for the various **above-code** recommendations in these terms permits homebuilders in any part of the nation to easily determine what the associated added cost will be in his or her area, thus allowing builders and potential homeowners to make reasonable cost-benefit decisions regarding implementation of the recommendations.

1.3 THE *INTERNATIONAL RESIDENTIAL CODE*

As already indicated, this guide focuses on the ICC's 2003 *International Residential Code*, which provides a comprehensive collection of requirements for prescriptive (nonengineered) residential construction. The *IRC*'s stated purpose is to provide:

> ... minimum requirements to safeguard the public safety, health, and general welfare, through affordability, structural strength, means of egress facilities, stability, sanitation, light and ventilation, energy conservation and safety to life and property from fire and other hazards attributed to the built environment.

The *IRC* addresses other natural hazards in addition to earthquakes (up to limits described in the *IRC* scoping provisions). When considering **above-code** recommendations, construction details intended to reduce the risk from one hazard may be slightly different from those needed to resist another. Thus, care should be taken to consider all natural hazards that present a risk to a specific site and to formulate an appropriate mitigation strategy in accordance with the jurisdiction's building code. Additional guidance is provided in Section 1.8 of this guide.

1.4 *IRC* SEISMIC DESIGN CATEGORIES

The *IRC* designates the level of potential seismic hazard for dwellings by assigning a house to a Seismic Design Category (SDC) based on its location. The *IRC* SDCs are A, B, C, D_1, D_2 and E, with A representing the lowest level of seismic risk applicable to residential construction and E, the highest. All residential buildings (detached houses and townhouses) in regions with SDC designations A and B, the lowest levels of seismic risk, are exempt from the seismic requirements of the *IRC*. SDC E regions have such a high level of seismic risk that, with a few exceptions, houses in these regions fall outside the scope of the *IRC* and must be designed using engineering principles following the *International Building Code* or *NFPA 5000*.

Whether or not required by the *IRC* and across all SDCs from A to E, many of the recommendations in this guide will improve the resistance of a dwelling to seismic forces, wind forces, and possibly the effects of other natural hazards. The discussion and examples presented in this guide focus on houses located in SDCs C, D_1, and D_2.

All U.S. model building codes provide maps identifying the seismic hazard. The 2003 *IRC* seismic design maps (*IRC* Figure R301.1(2) shown in Figure 1-1 designate the Seismic Design Categories for the nation and U.S. territories. It is a simplified version of the maps referenced by the *IBC* and *NFPA 5000* for all building types. The legend correlates the Seismic Design Category with the acceleration expected at a location in terms of gravity (g). A value of 100% g is equal to the vertical acceleration effects of gravity on Earth. More detailed information can be found for a particular building site using a CD-ROM available with the building codes and from FEMA; the CD allows the user to input the latitude and longitude coordinates of the site or the zip code. Zip codes should be used with caution because they may not reflect the highest possible SDC in the area covered by the postal zip code. Similar information is available on a U.S. Geological Survey (USGS) website – http://earthquake.usgs.gov/research/hazmaps (click on "seismic design values for buildings").

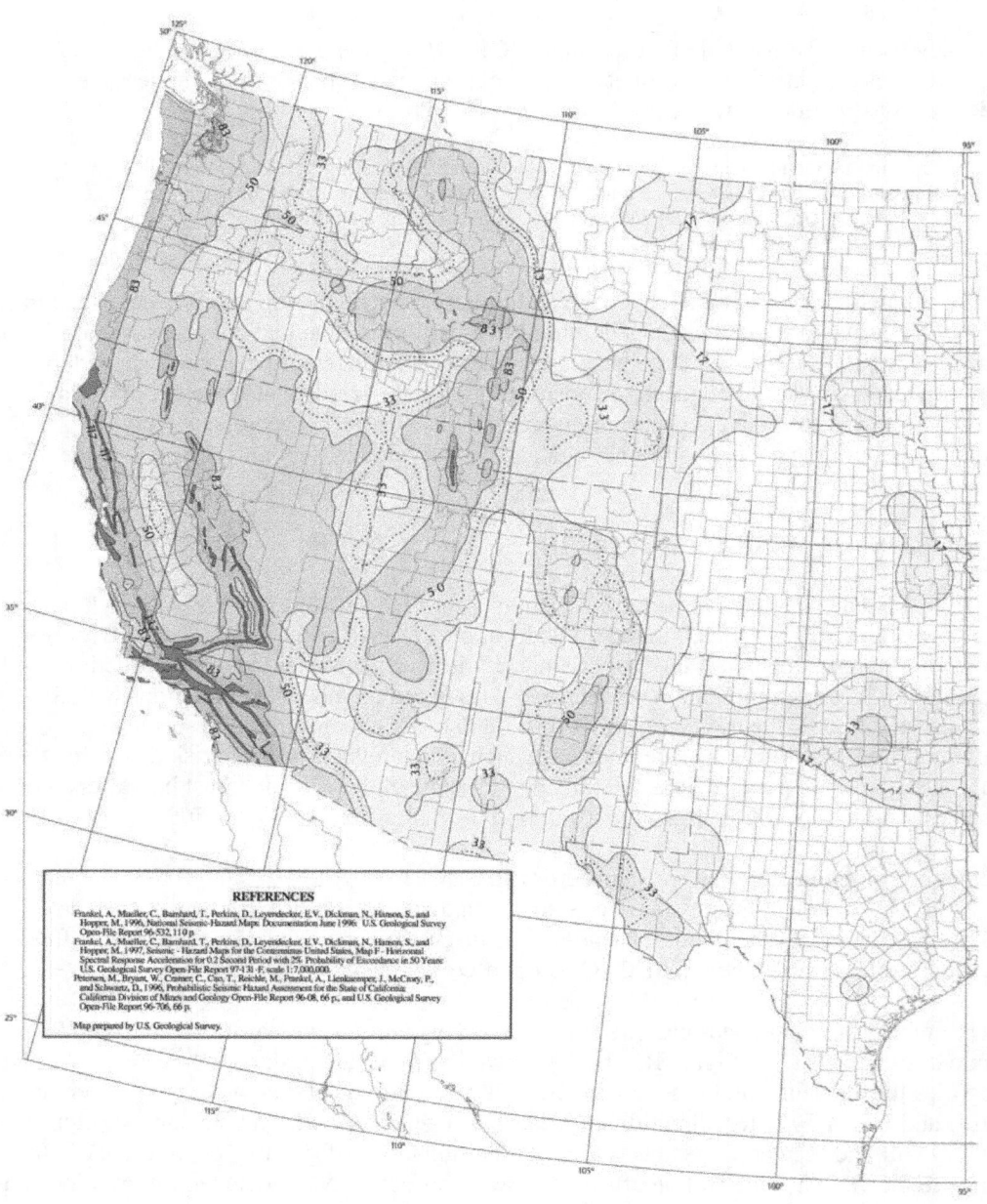

Figure 1-1 Seismic Design Categories – Site Class D.

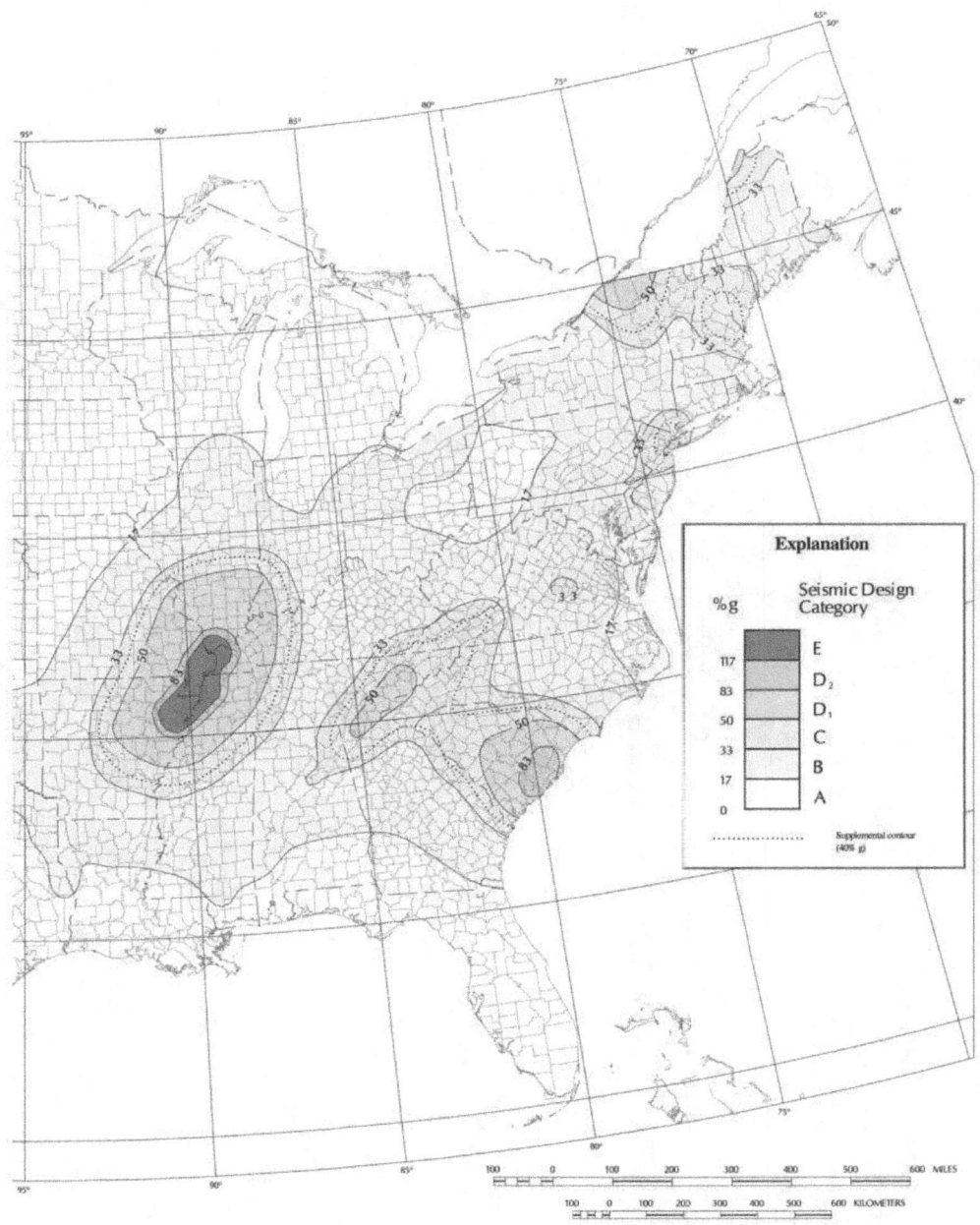

Figure 1-1 Seismic Design Categories – Site Class D (continued).

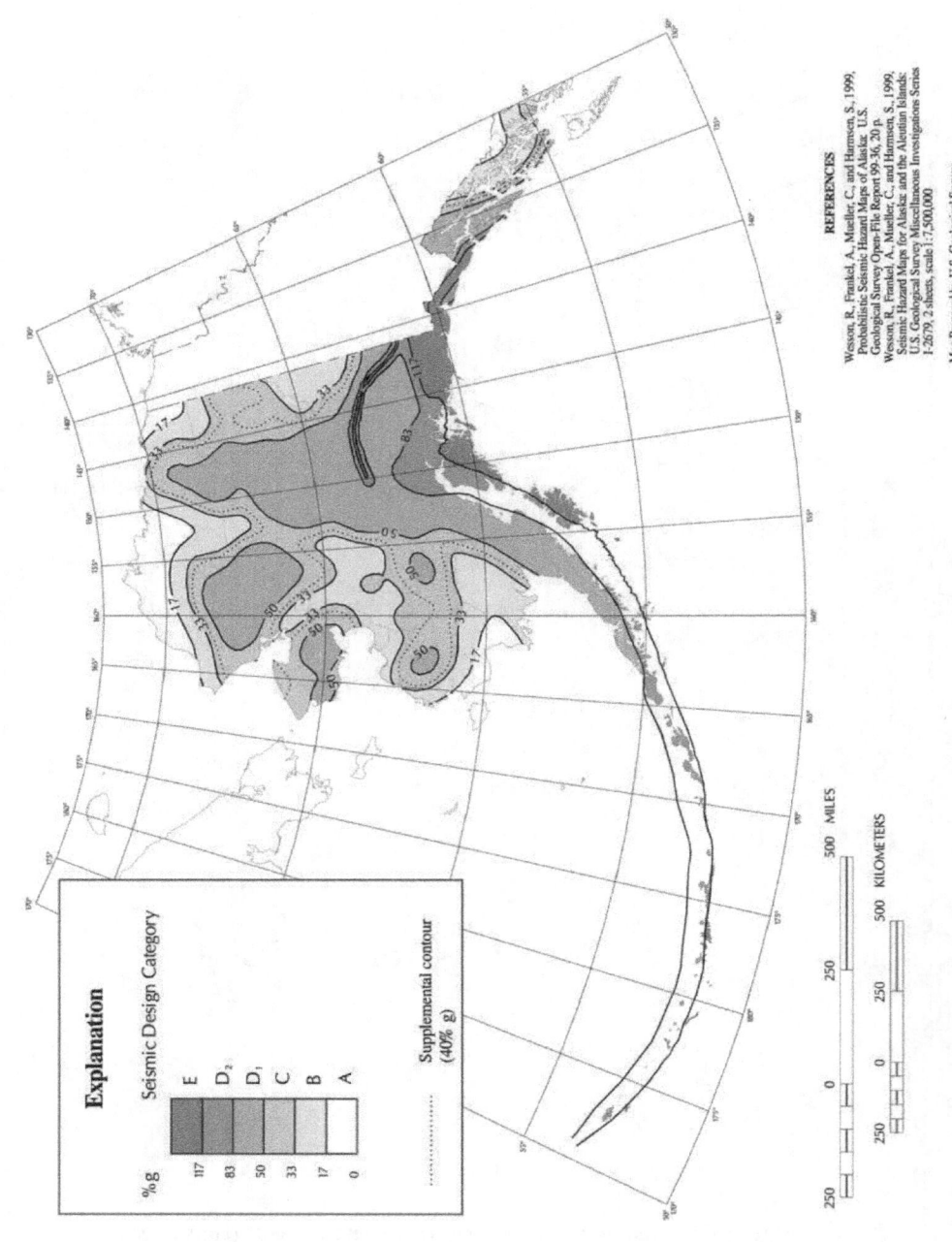

Figure 1-1 Seismic Design Categories – Site Class D (continued).

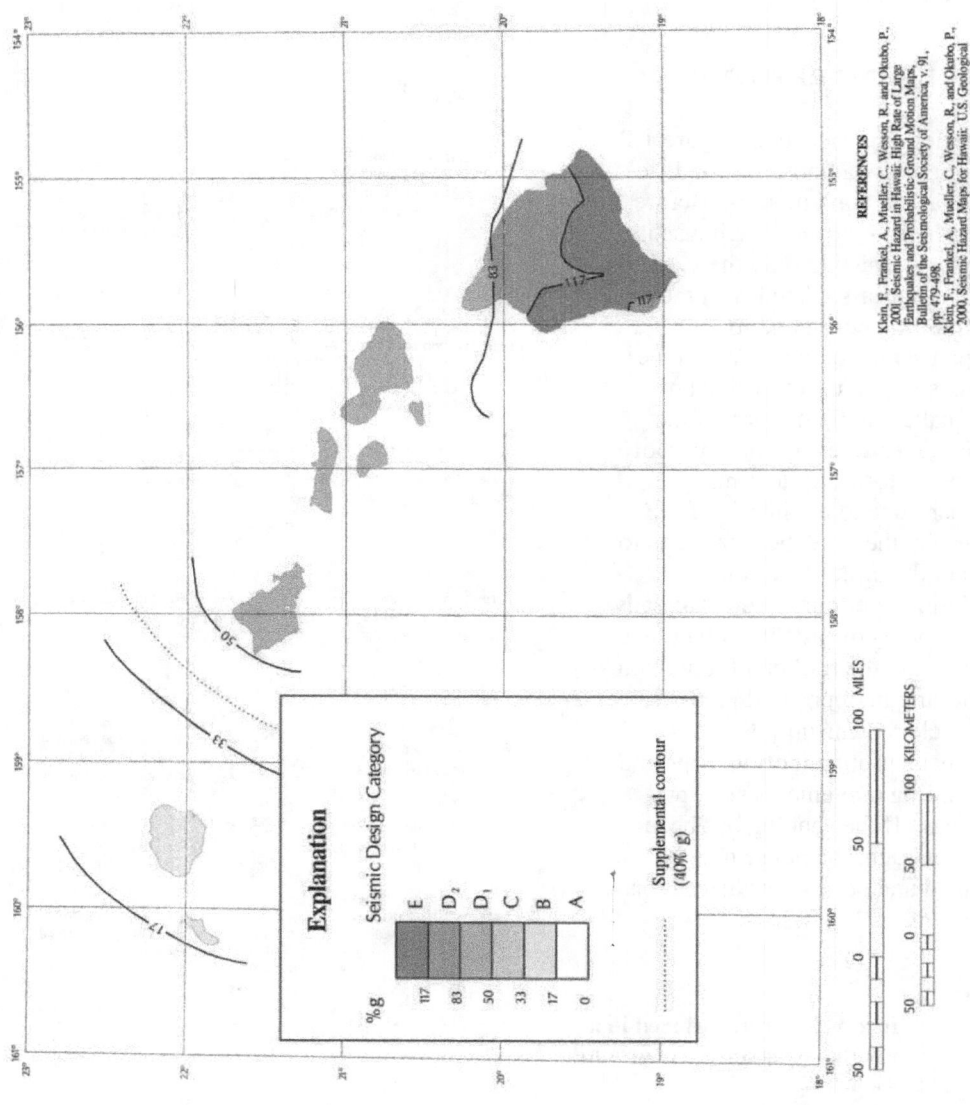

Figure 1-1 Seismic Design Categories – Site Class D (continued).

When using the seismic maps associated with the building codes, the reader should be aware that local soil conditions have a major impact on the seismic hazard for each particular site. The *IRC* map incorporates an assumed Site Class of D, which is stiff alluvial soil. If the soil conditions are different from stiff soil (e.g., bedrock or the soft alluvial soils often found in valleys), the local building department may have established special seismic requirements for these locations. Check with the local building department to determine if any such special regulations apply to a particular building site.

1.5 BUILDING RESPONSE

The series of drawings in Figure 1-2 illustrates how a house responds to earthquake ground motion. Before the earthquake occurs, the house is stationary, resisting only the vertical gravity loads associated with the weight of the house and its occupants and contents. When the ground starts to move during an earthquake, the foundation of the house moves sideways but the roof and upper stories try to remain stationary due to the inertia of the house. By the time the roof starts to move in the direction of the foundation, the foundation is already moving back towards its initial position and the roof and foundation are moving in opposite directions. This cycle repeats until the earthquake ground motion stops and the building movements slow and then stop. If the shaking has been severe enough, the house may be damaged and have a residual tilt or displacement.

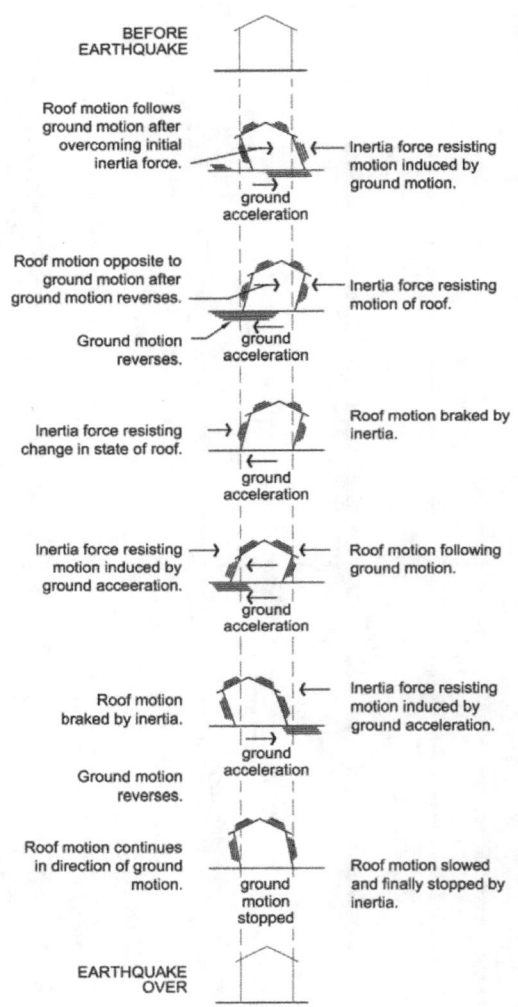

Figure 1-2 Forces induced in a house due to earthquake ground motion.

To perform adequately in an earthquake, a house must have enough strength to resist the forces generated without failure. Actual earthquakes can generate forces considerably higher than those used for code-prescribed design (Figure 1-3). Nevertheless, design for the lower code forces generally has prevented life loss and therefore satisfies the purpose of the code. Remember that the primary goal of the building code is to prevent loss of life; the prevention of damage is only incidental. It is for this reason that this guide presents **above-code** recommendations that describe techniques intended to improve the performance of a house during an earthquake and result in less damage.

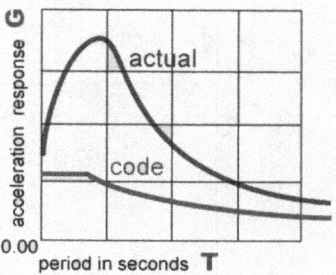

Figure 1-3 Concept of actual vs. code seismic forces.

Actual seismic forces may exceed the code design forces

Safety relies on contribution of nonstructural elements, excess material strength and ductility

The reasons one- and two-family houses tend to perform adequately in earthquakes even when designed to the minimum code forces are because houses often are stronger than recognized in code-level design and because they often are constructed with ductile earthquake-resisting systems. In residential construction, the finish materials and nonstructural partitions often add significantly to the strength provided by required bracing materials. Houses constructed from ductile earthquake-resisting systems generally will perform well during an earthquake because they can deform without breaking. An example of ductility is given in Figure 1-4. The ductile metal spoon simply bends while the brittle plastic spoon breaks.

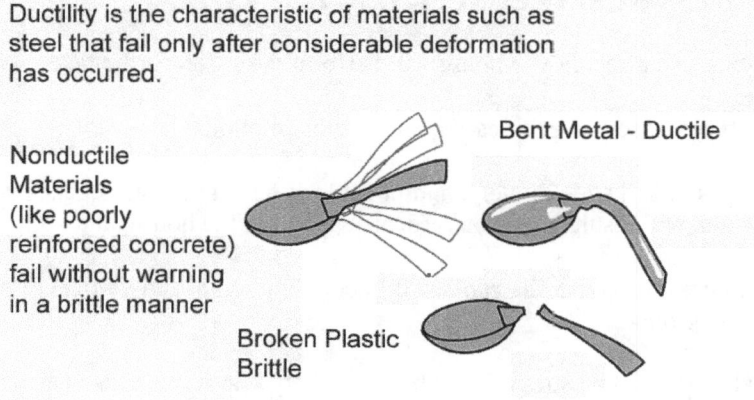

Figure 1-4 Concept of ductility.

Stiffness also is important. The stiffer the house, the less it will move or deflect during an earthquake, which will reduce the amount of damage to finish materials and, therefore, repair costs. Stiffness is measured in buildings in terms of drift (horizontal deflection) and is usually discussed as the horizontal deflection in a particular story in terms of either the amount of drift in inches or the percent story drift, which is the deflection in inches divided by the floor to floor story heights in inches (Figure 1-5). Unfortunately, increased stiffness also generally results in increased seismic forces; therefore, the **above-code** recommendations made in this guide simultaneously increase strength and stiffness.

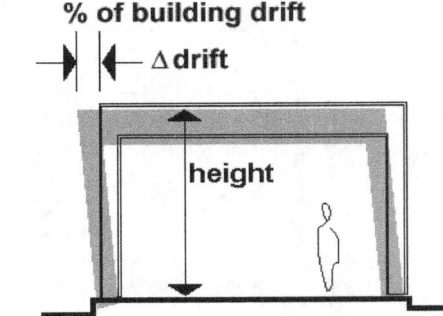

Figure 1-5 Definition of drift.

One way to understand how the strength and stiffness of a house affect its earthquake performance is to remember that the forces exerted by earthquake ground motion are resisted by a house's strength while the drift (deflection) is resisted by the house's stiffness. An analogy that can be used for this is a fishing pole. The amount the pole bends is dependent on how stiff the pole is; whether the pole breaks or not is dependent on how strong the pole is. A strong house will not fall down, but a house with relatively little stiffness will sustain considerable damage.

1.6 IMPORTANT CONCEPTS CONCERNING HOMESITE

Site characteristics also affect how a house will perform during an earthquake:

- Certain types of soil amplify earthquake ground motions;

- Some types of soil slide, liquefy, and/or settle as a result of earthquake ground motions, all of which will result in loss of vertical support for the house; and

- Fault rupture on a house site can result in both horizontal and vertical offsets of the supporting ground.

It is most desirable to build on sites with stable, solid geologic formations. Deep and unbroken rock formations, referred to as bedrock, generally will minimize earthquake damage whereas deep soft sedimentary soils will result in the maximum seismic forces and displacements being transferred to the house.

Sites located over known faults and in landslide-prone sites warrant additional attention. No matter how well designed, a house cannot accommodate earthquake ground motions at a site directly on top of an earthquake fault where the ground on either side of the fault moves in opposite directions. Check with the local building department to determine where known faults

Chapter 1, Introduction

are located. Houses should not be built within 50 feet of a known fault and, even at that distance, damage will still be significant. A house damaged by direct fault movement is shown in Figure 1-6.

Figure 1-6 A house damaged due to fault movement. The fault passed under this house near Wright's Station, on the Southern Pacific Railroad, in Santa Cruz County, California. The house was severed by fault movement during the earthquake. Photo credit: R.L. Humphrey, U.S. Geological Survey.

Sites where landslides are likely to occur also should be avoided. An example of damage due to a hillside collapsing in a landslide is shown in Figure 1-7. Engage a geotechnical or geological engineer to inspect the site and/or check with the local building department to minimize the risk of building on a landslide-prone site.

Figure 1-7 Houses damaged due to a landslide at the site.
Photo Courtesy National Information Service for Earthquake Engineering, University of California, Berkeley

FEMA 232, Homebuilders' Guide

1.7 MODEL HOUSE USED FOR GUIDE ANALYSIS

A typical model house is used in this guide to illustrate various concepts and techniques (Figure 1-8). An analysis of the model house identified approximate deflections under seismic loading, which permitted performance to be compared for various configurations using the minimum code requirements and the **above-code** techniques.

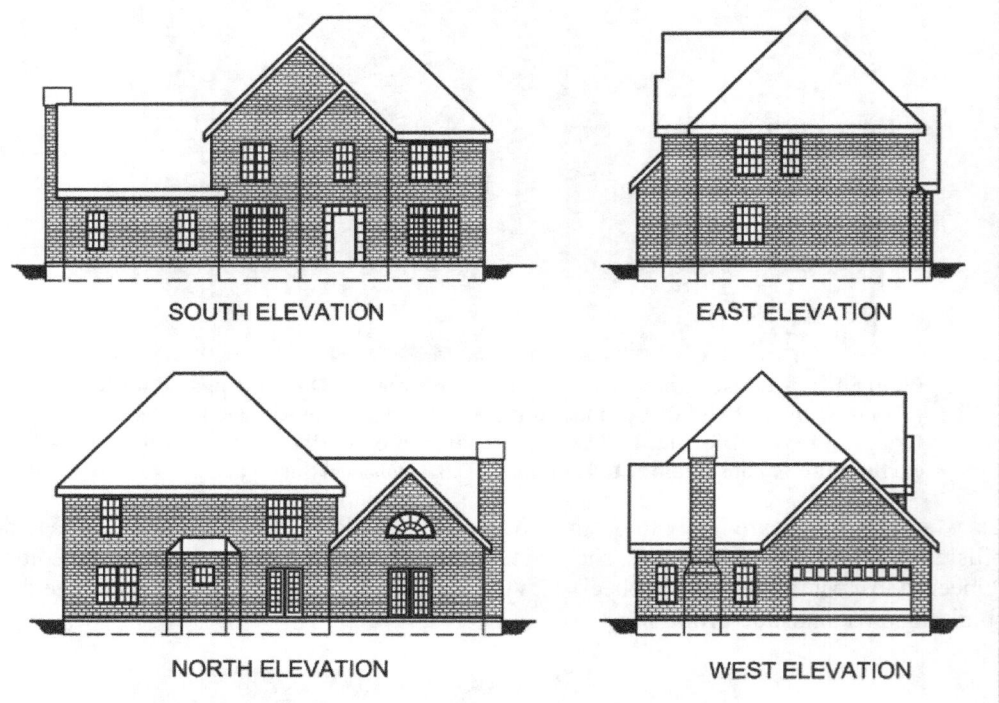

Figure 1-8 Elevations of the model house used as an example for this guide.

The floor plan of the model house is shown in Figure 1-9. *IRC*-conforming seismic bracing configurations were developed for SDCs C, D_1, and D_2. Gypsum wallboard was used to meet *IRC* seismic bracing requirements where possible because the wallboard would already be provided for wall finish. Where *IRC* bracing requirements could not be met with gypsum wallboard, wood structural panel (oriented strand board or OSB) wall bracing was added. Modifications to the basic floor plan included changing the foundation from a slab-on-grade to a crawl space (cripple wall or hillside condition) to illustrate the differences resulting from these changes.

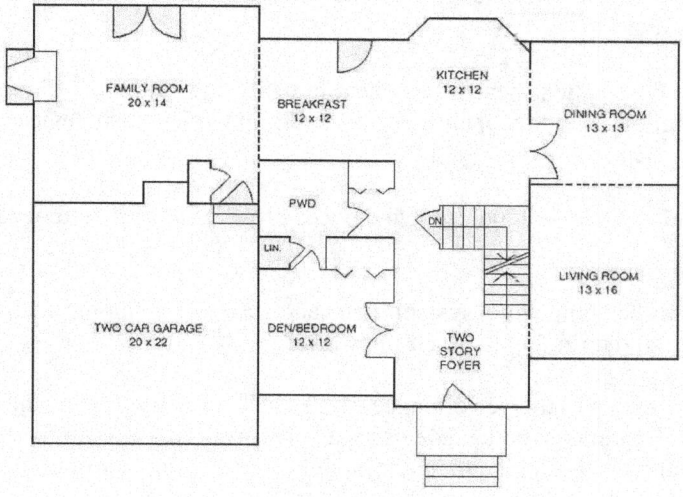

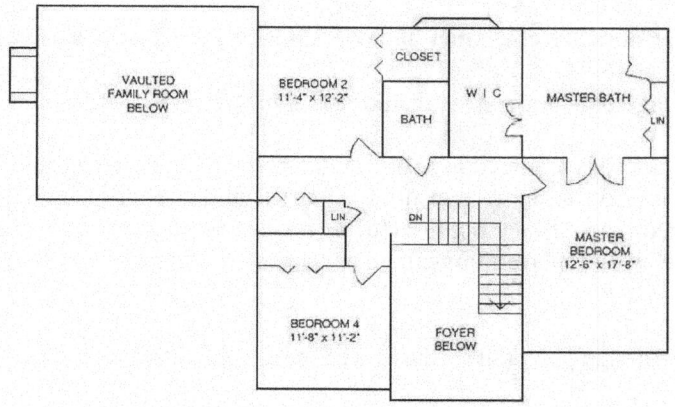

Figure 1-9 Model house floor plan.

After the basic model house construction was analyzed, the following **above-code** construction techniques were evaluated:

- A completely sheathed shear wall system in which the exterior walls were fully sheathed (around door and window openings) with OSB and hold-down devices were provided at building corners,

- A shear wall system with hold-down devices provided at each end of each OSB bracing wall, and

- An oversize sheathing panel system with each OSB sheet lapping vertically onto and edge-nailed to the rim joist at each floor level.

Approximate comparisons of the performance of the basic configuration of the model house with the **above-code** configurations were made using a two-part process. First, each house configuration was analyzed using a computer model to determine maximum drift or deformation during earthquake loading.

Second, in order to translate the analytical results into an approximation of house performance, three ranges of peak transient wall drift and assumed approximate descriptions of performance were developed. The choice of drift ranges and performance descriptions are based on component and full-building test results and the judgment of those participating in the development of this guide.

The approximate performance categories and corresponding story drift ranges are:

- **Minor** damage potential – Less than or equal to 0.5% story drift

    The house is assumed to suffer minor nonstructural damage such as cracking of plaster or gypsum wallboard and hopefully would be "green-tagged" (occupancy not limited) by inspectors after an earthquake, which would permit immediate occupancy. Some repairs should still be anticipated.

- **Moderate** damage potential – Above 0.5% to 1.5% story drift

    The house is assumed to suffer moderate damage including possible significant damage to materials and associated structural damage, but the building is assumed to have some reserve capacity in terms of strength and displacement capacity. The house hopefully would be "green-tagged" or, more likely, "yellow-tagged" (limited occupancy) by inspectors after an earthquake and may or may not be habitable. Significant repairs should be anticipated.

- **Significant** damage potential – Greater than 1.5% story drift

    The house is assumed to have significant structural and nonstructural damage that could result in its being "red-tagged" (occupancy prohibited) by inspectors after an earthquake. Significant repairs to most components of the building should be anticipated, and it may be more economical to replace the house rather than repair it.

Comparisons of performance based on this approach are discussed in a number of sections in this guide. See Appendix A for a detailed description of the analysis.

1.8 OTHER HAZARDS

Although this guide has been written primarily to address the earthquake risk to a house, homebuilders and homeowners need to be aware of many other natural hazards. These hazards can also affect how a house should be designed and constructed as well where it should be sited and even whether the site should be used at all.

1.8.1 Wind

Since wind acts on a house by imparting horizontal loads similar to those of an earthquake, the seismic design and construction criteria contained in the *IRC* and further detailed in this guide will provide a significant level of protection to a house's structural system against wind loads. However, wind loads also act on the home's cladding, walls, and roof coverings, and the wind provisions of the *IRC* should be followed carefully (due to the differences in the load path for wind versus earthquake loads). While the *IRC* wind provisions are adequate, additional guidance is available and should be considered. This guidance includes the American Forest and Paper Association (AF&PA) *Wood Frame Construction Manual* and the Southern Building Code Congress International (SBCCI) *Standard for Hurricane-Resistant Residential Construction* (SSTD 10).

1.8.2 Flooding

Care should be taken to ensure that the home site is not in a flood hazard area. If there is any doubt in this regard, the local building code enforcement or floodplain management office should be consulted. As part of the National Flood Insurance Program (NFIP), FEMA publishes local flood hazard maps that delineate the extent of the flood hazard, using the Base Flood Elevation or BFE, which is flood elevation that has a 1 percent chance of occurring in any given year. As part of its agreement to participate in the NFIP in order to make federally-backed flood insurance available to its residents, a community adopts and enforces a series of flood design and construction requirements. Generally, these requirements mean that a home located in a flood hazard area must have its lowest floor and utilities elevated to or above the BFE.

FEMA has available a series of publications on protecting homes from flood damage. Some of these publications are:

- *Design Guidelines for Flood Damage Reduction* (FEMA 15)
- *Elevated Residential Structures* (FEMA 54)
- *Above the Flood: Elevating Your Flood-prone House* (FEMA 347)

For flood protection of existing houses, the reader is referred to *Homeowner's Guide to Retrofitting: Six Ways to Protect Your House from Flooding* (FEMA 312). All of the above publications are available from the FEMA Publications Center at 1-800-480-2520. Additional information is also available online at www.fema.gov.

For guidance on the flood resistant construction provisions of the *IRC*, the reader is referred to the document *Reducing Flood Losses Through the International Codes: Meeting the Requirements of the National Flood Insurance Program*. This publication is available online at www.fema.gov as well as for a modest fee from the International Code Council (www.iccsafe.org) at (800) 786-4452.

Above-code Recommendations: If a house is to be located in a flood hazard area and must be elevated to the BFE, it is recommended that the lowest floor of the house actually be elevated 2 or 3 feet above the BFE. This will result in a significantly lower insurance premium for the homeowner and will provide a significantly higher level of protection for a minimal additional cost.

If a house is to be located just outside of a flood hazard area and is not required to meet a community's flood regulations, it is recommended that the home still meet the community's flood ordinances, such as not using a basement foundation, to provide an additional level of protection. Further, the homeowner is still encouraged to purchase flood insurance. Even if a home is just outside of a flood hazard area, it can still be subject to flooding as higher floods than the design flood can and have occurred. As a further measure of protection, flood insurance is available outside of the flood hazard area, usually for a very reasonable premium.

1.8.3 Coastal Hazards

Homesites along the coast are subject to two different hazards from hurricanes and other coastal storms such as Northeasters — high winds and flooding from storm surge. Storm surge is caused by the storm's low pressure and winds pushing the water onto land. In the recent Hurricane Katrina, storm surge heights of almost 30 feet were measured along the coast of Mississippi. Coastal flood hazard maps similar to the flood maps described above are prepared by FEMA; however, the coastal flood hazard area is divided into two zones, the coastal high hazard zone or "V-Zone" which has significant wave heights and the regular flood hazard zone or "A-Zone" which does not.

Coastal homes must be designed and built to withstand both wind and storm surge loads. This requires elevating the house above mapped storm surge elevations and ensuring that the house can resist the loads associated with hurricane force winds. Further, since storm surge waters can flow with considerable velocity which generates substantial hydrodynamic loads on the area below the elevated floor, this area must be built to allow water to flow through without placing additional loads on the structure. For this reason, any walls below the lowest floor must be made of "breakaway" construction.

Note also that the coastline is a dynamic and ever-changing environment and is often subject to erosion. In choosing a home site, consider whether the property is subject to erosion, which could easily undermine and destroy the house in a relatively short period of time. Erosion maps showing the rate of change of the shoreline may be available locally.

Above-code Recommendations: For additional protection of coastal homes, the following is recommended:

- **Houses should be elevated 2 or 3 feet above the storm surge flood elevation provided on local maps.**

- **The area below the lowest floor should be kept open and free of as many obstructions as possible.**

- **Houses located in A-Zones but still near the coast (also known as Coastal A-Zones) may still be subject to waves and storm surge and should be constructed to V-Zone criteria.**

- **Houses should be designed to resist damage from wind and debris. This includes limiting the size and amount of glazing and actually installing the storm shutters before an event.**

- **Building sites should be investigated for erosion potential prior to purchase and sites subject to erosion should be avoided. Houses built where erosion is possible should be located as far from the shoreline as possible.**

- **Natural features such a sand dunes and vegetation should be used as much as possible to protect houses from coastal forces.**

All of the above recommendations will provide better protection than currently offered by the *IRC* with little additional cost.

The reader is encouraged to consult the *FEMA Home Builders Guide to Coastal Construction: Technical Fact Sheet Series* (FEMA 499). These fact sheets contain design and construction suggestions based on lessons learned in FEMA's investigations of past hurricanes and how they impacted residential construction. This document is available from the FEMA Publications

Center as well as online at www.fema.gov. For more detailed engineering design information, the reader may wish to consult the *Coastal Construction Manual: Principles and Practices of Planning, Siting, Designing, Constructing, and Maintaining Residential Buildings in Coastal Areas* (FEMA 55). This is a large document (500 pages) and is available from the FEMA Publications Center in paper or on a CD-ROM.

1.8.4 Tornado

Building codes generally do not contain tornado provisions primarily because the probability of a specific house being hit by a tornado powerful enough to do damage are sufficiently remote that it is not thought to be economically justifiable to design for one. Therefore, further guidance on resisting the loads from a tornado does not exist. If protection of a home's occupants is a concern, the reader is referred to *Taking Shelter from the Storm: Building a Safe Room Inside Your House* (FEMA 320). This document is available from the FEMA Publications Center as well as online at www.fema.gov.

Chapter 2
EARTHQUAKE-RESISTANCE REQUIREMENTS

This chapter explains the *International Residential Code*'s (*IRC's*) general earthquake-resistance requirements as well as specific *IRC* requirements concerning load path and house configuration irregularities. One- and two-family detached houses of wood light-frame construction are addressed; however, the discussion is relevant to other materials of construction likely to be used for detached houses including light-frame cold-formed steel.

2.1 *IRC* GENERAL EARTHQUAKE LIMITATIONS

The variety of configurations used for houses is very wide and they are constructed of an equally wide variety of materials. *IRC* Section R301.2.2 imposes some limits on configuration and materials of construction for one- and two-family detached houses in Seismic Design Categories (SDCs) D_1 and D_2. These *IRC* limitations reflect the desire to provide equal earthquake performance for houses designed using the prescriptive *IRC* provisions and for those with an engineered design. Application of the prescriptive *IRC* requirements to houses that do not comply with the limitations can be expected to result in inadequate performance.

IRC Section R301.2.2 earthquake limitations do not apply to one- and two-family detached houses in SDCs A, B, and C; however, townhouses in SDC C are required to comply. In general, design using the *IRC* is prohibited in SDC E, and the designer is referred to the engineered design requirements of the *International Building Code* (*IBC*) or other adopted code. There are, however, two methods by which houses designated as SDC E in *IRC* Figure R301.2(2) may be designed in accordance with *IRC* provisions: (1) if site soil conditions are known, *IRC* Section R301.2.2.1.1 permits determination of the Seismic Design Category in accordance with the *IBC*, which may result in a reduced SDC, and (2) if the restrictions of *IRC* Section R301.2.2.1.2 are met, houses located in SDC E may be reclassified to SDC D_2 and designed using the *IRC* provisions. *IRC* Section R301.2.2.1.2 requirements dictate a regular house configuration.

The limitations imposed by *IRC* Section R301.2.2 are as follows:

- *Weight Limitations* – For houses in SDCs D_1 and D_2, *IRC* Section R301.2.2.2.1 specifies maximum weights for the floor, roof-ceiling, and wall assemblies or systems. Because earthquake loads are proportional to the weight of the house, an upper bound on assembly weight provides an upper bound on earthquake loads. The specified maximum assembly weights relate directly to the weights considered in developing the *IRC* earthquake bracing provisions. The effect of the maximum weights is the exclusion of heavier finish materials when using the *IRC* provisions. Where heavier finish materials are to be used, an engineered design must be provided.

- *House Configuration Limitations* – For houses in SDCs D_1 and D_2, *IRC* Section R301.2.2.2.2 places limits on house configuration irregularities. These limits are discussed in detail in Section 2.3 of this guide.

- *House System Limitations* – Another scope limitation for houses in SDCs D_1 and D_2 is given in the combined requirements of *IRC* Sections R301.2.2.3 and R301.2.2.4. These sections provide limits for number of stories based on building system and limits for anchored stone and masonry veneer and masonry and concrete wall construction. Some of these requirements are addressed in Chapter 5 of this guide.

- *Story Height Limitation* – *IRC* Section R301.3 provides a scope limitation that is not related solely to earthquake loads but rather applies in all SDCs. This section limits story height by limiting the wall clear height and the height of the floor assembly. This limits both the lateral earthquake and wind loads and the resulting overturning loads.

The *IRC* requires design in accordance with accepted engineering practice when the general earthquake limitations discussed above are not met (weight limitations, house configuration limitations, building system limitations, and story height limitations). Engineered design is addressed in Section R301.1.3. This section permits design to be limited to just the elements that do not conform to the *IRC* limitations. Increased assembly weight and story height will globally increase seismic and wind loads, generally making engineered design of the entire house necessary. Design of portions of the house is particularly applicable when an irregularity such as a cantilever, setback, or open front occurs. The extent of design is left to the judgment of the designer and building code official. When multiple irregularities occur, engineered design of the entire house may become necessary in order to provide adequate performance. The *IRC* requires that engineering design methods be used but does not specify whether this must be done by a registered design professional. State or local law governs who can perform the design; the reader is advised to check with the local building department for requirements.

2.2 LOAD PATH

For a house to remain stable, a load applied at any point on the structure must have a path allowing load transfer through each building part down to the building foundation and supporting soils. The term "load path" is used to describe this transfer of load through the building systems (floors, roof-ceilings, bracing walls).

Basic Concept — To understand the concept of a load path, a house can be represented by the chain shown in Figure 2-1. The chain is pulled at the top and the load is transferred from one link to the next until it is transferred to the ground. If any link is weak or missing, the chain will not adequately transfer the load to the ground and failure will result.

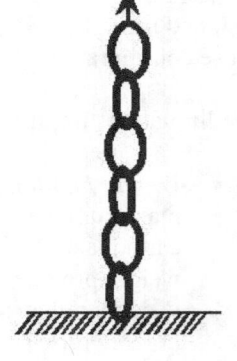

Figure 2-1 Chain illustrating the load path concept.

Likewise, houses must have complete and adequate load paths to successfully transfer earthquake loads and other imposed loads to the supporting soils.

Load Path for Earthquake and Wind Loads — The example house in Figure 2-2 will be used to discuss load path. The arrows provide a simplified depiction of earthquake or wind loads pushing horizontally on the house. Although wind and earthquake loads can occur in any horizontal direction, design procedures generally apply the loads in each of the two principal building directions (i.e., longitudinal and transverse), one at a time, and this discussion of loading will utilize that convention.

Internally, the house has to convey loads from the upper portions of the structure to the foundation. For the example house, this is accomplished by transferring the loads through:

- The roof-ceiling system and its connections to the second-story bracing wall system,
- The second-story bracing wall system and its connections to the floor-ceiling system,
- The floor-ceiling system and its connections to the first-floor bracing wall system, and
- The first-story bracing wall system and its connections to the foundation, and
- The foundation to the supporting soil.

The following discussion focuses primarily on the connections between the various building systems; the systems themselves are addressed only briefly here but are discussed in detail in later chapters of this guide.

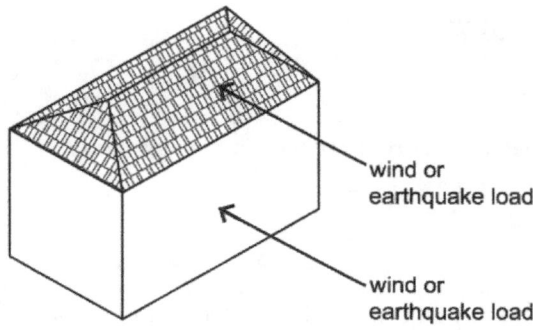

Figure 2-2 Lateral loads induced in a building due to wind or earthquakes.

Roof-ceiling and Floor Systems — In the example house, the roof-ceiling system will resist horizontal earthquake loads proportional to the weight of the roof, ceiling, and top half of the second-story walls. The series of arrows at the right of Figure 2-3a depicts this load. The roof-ceiling system deflects horizontally under the load and transfers the load to the supporting walls at both ends. The single arrows at the roof-ceiling system ends depict the reaction loads to the supporting walls. Within the roof-ceiling system, the load is carried primarily by the roof sheathing and its fastening (see Chapter 6 of this guide for discussion of roof-ceiling systems).

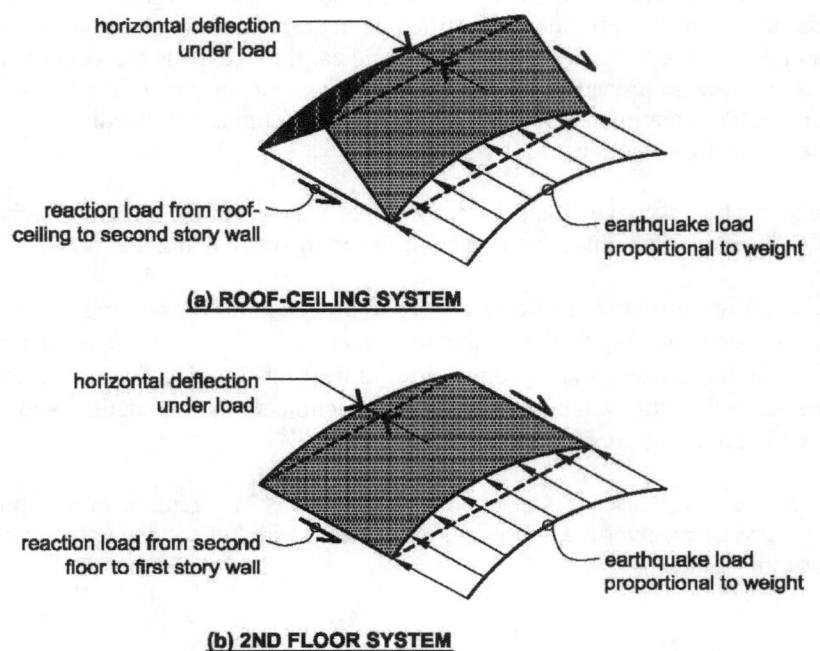

Figure 2-3 Loading and deflection of roof-ceiling and floor systems.

Similarly, the floor system will resist horizontal earthquake loads proportional to its weight and the weight of walls above and below. As shown in Figure 2-4b, it will deflect and transfer load to the supporting walls in much the same way as the roof-ceiling system. Again, the loading is carried by the floor sheathing and its fastening (see Chapter 4 of this guide for discussion of floor systems).

Bracing Wall Systems – The roof-ceiling reaction load is transferred into the second-story bracing wall system as depicted by the arrow at the top of the wall in Figure 2-4a. The wall deflects under this load and transmits the load to the wall base and through the floor system to the first-story wall. Resistance to the wall load is provided by the wall sheathing and its fastening.

Chapter 2, Earthquake-Resistance Requirements

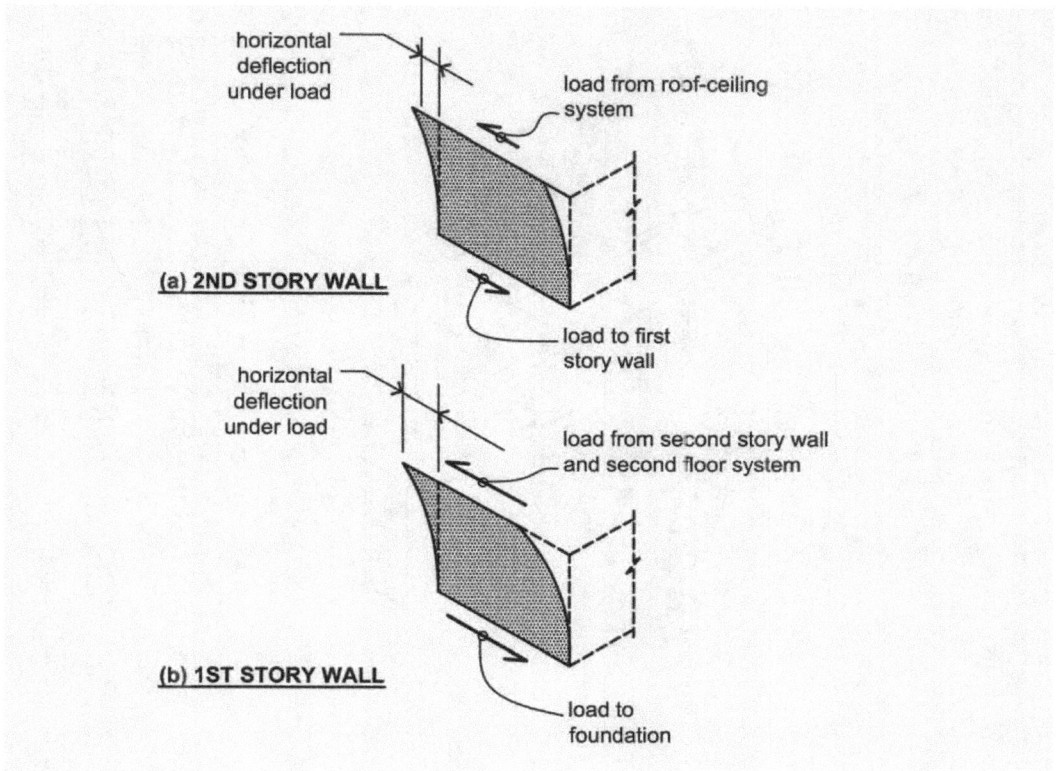

Figure 2-4 Loading and deflection of bracing wall systems.

The first-story bracing wall system resists loads from both the second-story wall and the second-story floor system as depicted by the arrow at the top of the wall in Figure 2-4b. The wall deflects under this load and transmits the load to the wall base and the foundation. Again, resistance to the wall load is provided by sheathing and its fastening. Figure 2-5 provides an exploded view of the example house that illustrates the combination of roof-ceiling, floor, and wall systems and their connection to the foundation below.

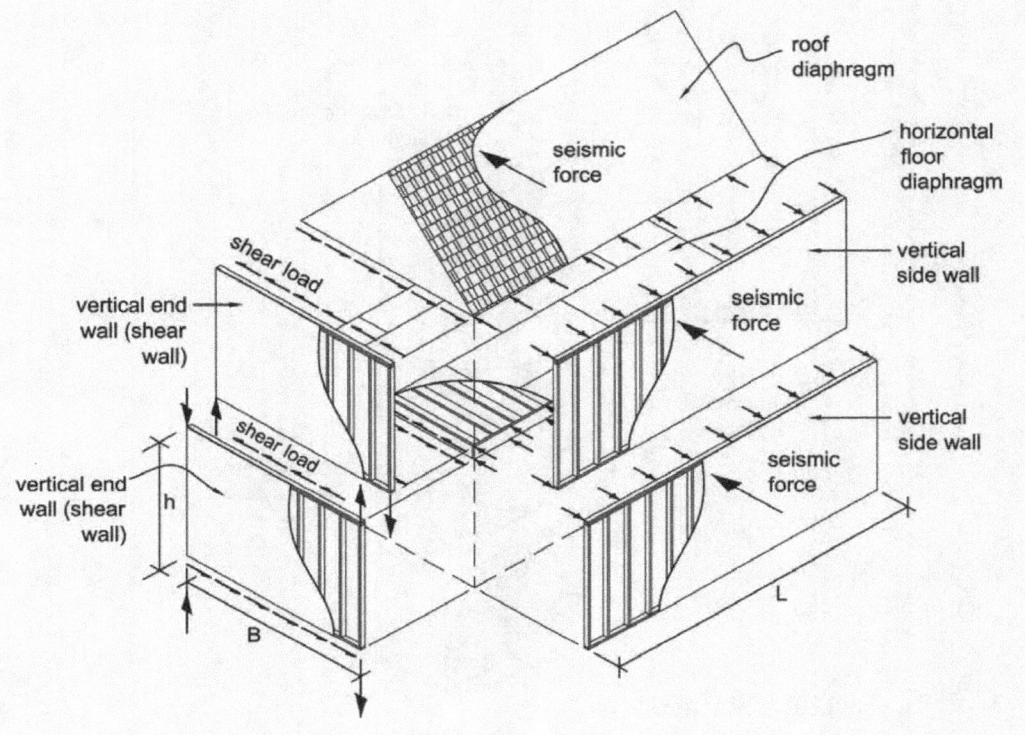

Figure 2-5 Load transfer between components in a building.

Requirements for Connections Between Systems – As previously noted, a complete load path for earthquake loads requires not only adequate roof-ceiling, floor, and bracing wall systems but also adequate connection between these systems. Connections between systems must resist two primary types of loads: horizontal sliding loads and overturning loads.

Load Path Connection for Horizontal Sliding – Figure 2-6 depicts the end wall at the left side of the house illustrated in Figures 2-2 through 2-5 and provides a detailed illustration of one possible path for horizontal loads from the roof assembly to the foundation. The left-hand portion of the figure shows a section through the end wall in which each of the "links" in the load path is given a number, H1 through H11, corresponding to a connection or mechanism used to transfer the loads. The right-hand side of the figure shows an elevation of the same wall and illustrates the deformation that will occur if adequate connection is not made. Table 2-1 at the end of this chapter provides a detailed summary of the load path connections for the Figure 2-6 wall.

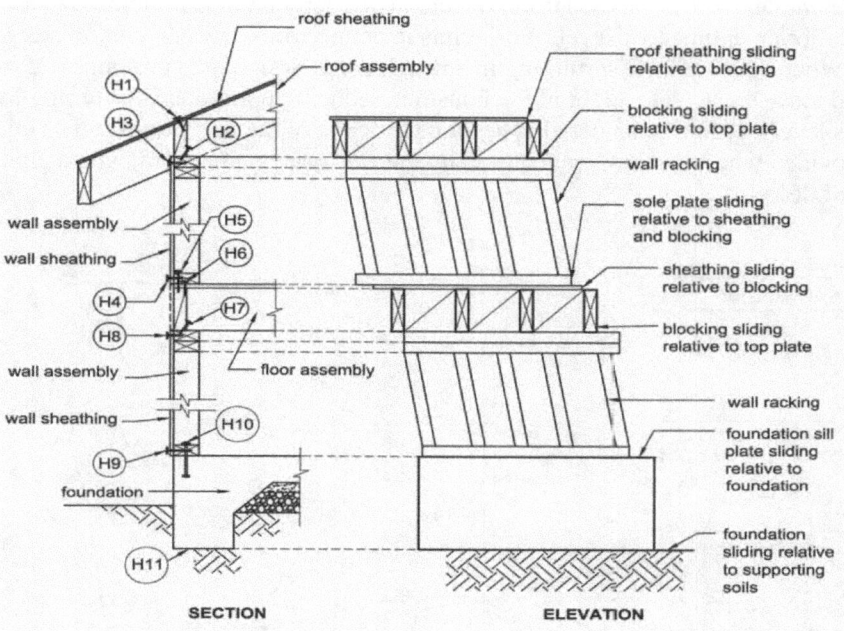

Figure 2-6 Horizontal load path connections and deformations.

Load Path Connection for Overturning – Because the horizontal loads are applied high on the house and resisted at the foundation, overturning loads develop in the bracing walls. Figure 2-7 illustrates one possible load path for overturning loads. The left-hand side of the figure shows a wall elevation in which each of the "links" in the overturning load path is given a number, OV1 to OV8, corresponding to locations with uplift or downward loads due to overturning. The right-hand side of the diagram shows an elevation of the same wall that illustrates overturning deformations that will occur with earthquake loading from the left to the right. Uplift or tension occurs at one end of a wall simultaneously with downward force or compression at the other end.

The *IRC* only specifies connections (hold-down straps or brackets) to resist overturning loads for a limited number of bracing alternatives. The connectors used to resist horizontal loads in most *IRC* designs will be required to resist overturning loads as well as horizontal loads. This is a major difference between prescriptive design and engineered design in which resistance to overturning loads must be explicitly provided. The *IRC* requires use of hold-down devices in Section R602.10.6 for alternative braced wall panels; in Section R602.10.11, Exception 2, for SDCs D_1 and D_2 when braced wall panels are not located at the end of braced wall lines; and in Section R703.7 when stone or masonry veneer is used. Table 2-2 at the end of this chapter provides a detailed summary of the load path connections for the Figure 2-7 wall.

Figure 2-7 graphically depicts use of straps to resist overturning loads; however, overturning can be resisted by connections employing such other devices as bolts, nails, or hold-down brackets. Because different device types may deform differently under load, it is preferable to use the same type of device for an entire story level. Variations in connector type from story to story are acceptable. When considering overturning in an engineered design, it is customary to include the effect of dead load (the actual weight of the house) in reducing uplift and overturning loads; however, this level of calculation detail is beyond the scope of the *IRC* provisions. Hold-downs should be provided wherever they are required by the *IRC* or recommended by this guide, irrespective of dead load.

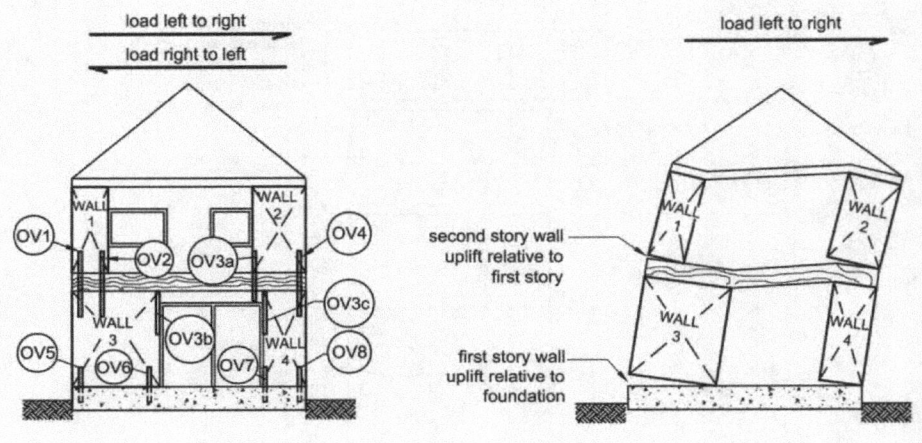

Figure 2-7 Overturning load path connections and deformations.

System and Connection Strength and Stiffness Requirements – The roof-ceiling, floor, and bracing wall systems are the basic members resisting earthquake loads. Adequate earthquake performance of a house relies on:

- Adequate strength of roof-ceiling, floor, and bracing wall systems,
- Adequate stiffness of roof-ceiling, floor, and bracing wall systems to limit deformation,
- Adequate connection between systems to provide a functional load path, and
- Adequate connection to the foundation.

For most houses, it is generally anticipated that bracing wall system behavior will have more influence on the behavior of the house than the roof-ceiling or floor systems. Further, it is anticipated that wall behavior at lower stories will generally be more critical than at upper stories due to larger earthquake loads.

Accumulation of Loads in Systems and Connections – Wind and earthquake loads increase or accumulate towards the bottom of the house. This is particularly applicable to loads in the bracing wall systems and their connections for horizontal loads and uplift and downward loads due to wall overturning. For example, overturning connections must be sized to resist all of the loads generated above the connection location. In a two-story house, the second-floor uplift connection, such as OV1 in Figure 2-7, will need to resist loads from the second story. The first-story uplift connection, such as OV5 in Figure 2-7, will have to resist the uplift loads from the second story plus the additional uplift from the first story. It can generally be expected that OV5 will need to resist a load two to three times that resisted by OV1. Downward loads at the opposite ends of walls and horizontal loads accumulate similarly. Although this accumulation of load often is missed in design for wind and earthquake loads, it is important and should be explicitly considered when connections are being selected.

> **Above-code Recommendation: For overturning loads, the use of hold-down connectors (brackets, straps, etc.) is recommended as an above-code measure in order to improve load transfer and thereby decrease damage.** For most of the basic *IRC* wall bracing types, the provisions rely on building weight and the fasteners resisting horizontal loads to also resist overturning. Wall component testing, however, indicates that this method is not as reliable in resisting overturning as the use of positive anchorage devices.

2.3 HOUSE CONFIGURATION IRREGULARITIES

A house's configuration (shape) significantly affects its response to wind and earthquake loads. This section discusses the concepts and the *IRC* provisions for irregular house configurations.

Ideal Earthquake-Resistant House Configuration – For earthquake resistance, the ideal house would have:

- A simple rectangular shape,
- Bracing walls distributed uniformly and symmetrically through the house,
- No large concentrations of weight,
- Bracing walls at upper stories located immediately above walls in stories below,
- Wall bracing lengths that increase in lower story levels compared to the story above,
- No split-levels or other floor level offsets.

A version of this ideal house is shown in Figure 2-8. This ideal configuration results in loads and deformations being uniformly distributed throughout the house, which permits resisting elements to contribute equally to earthquake resistance. With good distribution of bracing walls, earthquake loads can be resisted very close to where they are generated (by house weight), which reduces the need for transfer of earthquake loads through floor and roof systems to other portions of the house. This helps reduce the poor performance that often results when such transfers are required.

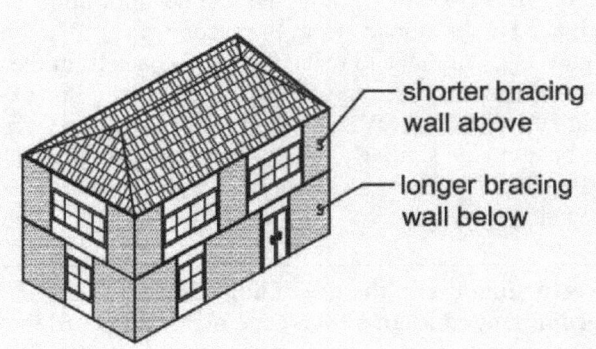

Figure 2-8 Ideal building configuration for earthquake resistance.

Deviations from Ideal Configuration – Deviations from the ideal configuration are called configuration irregularities. As houses deviate from the ideal configuration, loads and deformations are concentrated, which causes localized damage that can result in premature local or even complete failure of the house. While the ideal configuration is attractive from the standpoint of earthquake resistance, houses with irregularities are much more popular and common than those without. Large open great rooms and walls of nothing but windows are examples of typical irregular configurations.

House Irregularity Concepts – House irregularities often are divided into two types: plan irregularities and vertical irregularities.

Plan irregularities concentrate earthquake load and deformation in a particular area of a house. A common cause is a center of mass (building weight) at a location different from the center of the resisting elements (bracing walls). This can be due to non-uniform mass distribution, non-uniform distribution of bracing walls, or an irregular house plan. Two common examples are a house with one exterior wall completely filled with windows with no wall bracing provided (Figure 2-9) and a house with a large masonry chimney at one end. Houses with plan irregularities generally experience rotation in addition to the expected horizontal deformation. House rotation, as illustrated in Figure 2-10, magnifies the displacements, resulting in increased damage.

Chapter 2, Earthquake-Resistance Requirements

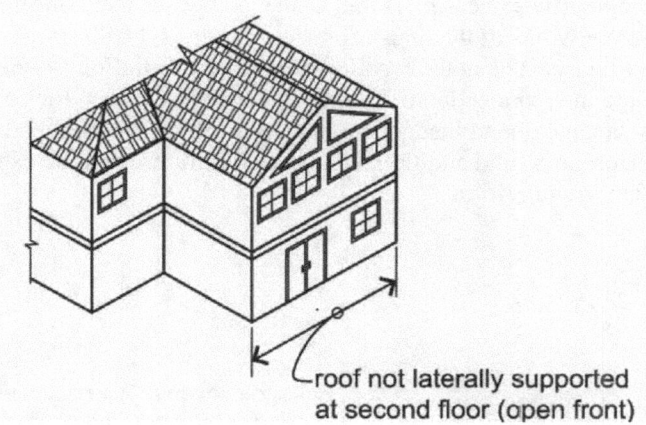

Figure 2-9 Irregular building configuration with open front (no second-story bracing walls supporting roof on right-hand portion).

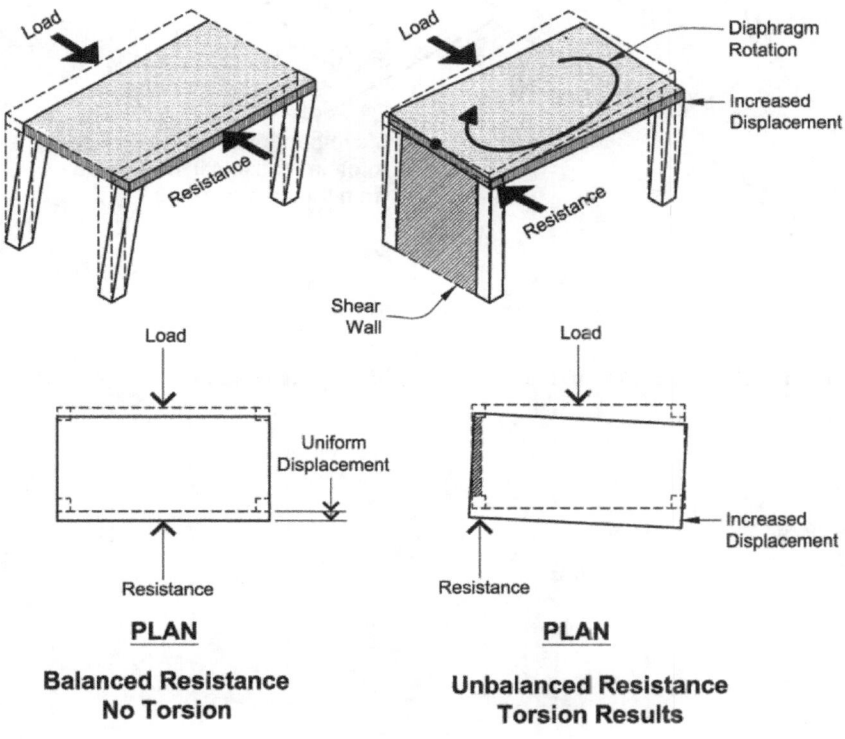

Figure 2-10 Rotational response and resistance to torsion.

FEMA 232, Homebuilder's Guide

Other common plan irregularities occur in T- and L-shaped houses that concentrate loads at the corners where the different wings of the house connect. Figure 2-11 illustrates the concentration of loads in an L-shaped house. The noted location of load concentration is where damage would be anticipated. Adequate interconnection of the house wings is required for good performance. Figure 2-12 illustrates damage due to inadequate interconnections. Without additional consideration, poor performance and additional damage or failure would be expected for structures with such plan irregularities.

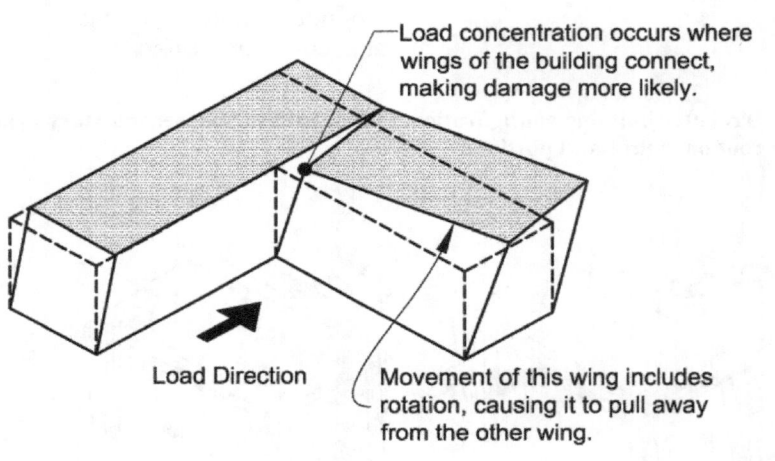

Figure 2-11 Concentration of loads and possible damage and failure due to irregular plan.

Figure 2-12 Roof-level damage to building with a T-shaped plan. The damage at the roof corner (right-hand side) is due to stress concentrations where the wings of the building connect.

Vertical Irregularities – Vertical irregularities concentrate damage in one story of a multistory house. This occurs when the stiffness or strength of any one story is significantly lower than that of adjacent stories. When the stiffness of a story is significantly lower, the deformations associated with the earthquake loads tend to be concentrated at that "soft" story as illustrated in Figure 2-13. The configuration on the left (normal) provides a uniform stiffness in each story, while the configuration in the middle illustrates a soft story with the deflections being concentrated in that story. If the displacements get large enough, they can cause complete failure of the soft story as illustrated in the configuration on the right.

The CUREE-Caltech Woodframe Project found that many two-story light-frame residential houses exhibit soft-story behavior to some extent because the first stories feature relatively large window and door openings and fewer partitions (less bracing) than the second stories where the bedrooms and bathrooms are located. Soft first-story behavior also was observed in the analysis of the model house used for this guide. Figure 2-14 provides an exaggerated illustration of deformation concentrated in the first story of the model house. Cripple walls around the perimeter of a crawlspace also can result in soft-story behavior. A weak story can cause damage or failure to be concentrated in that story if earthquake loading approaches the story strength.

FEMA 232, Homebuilder's Guide

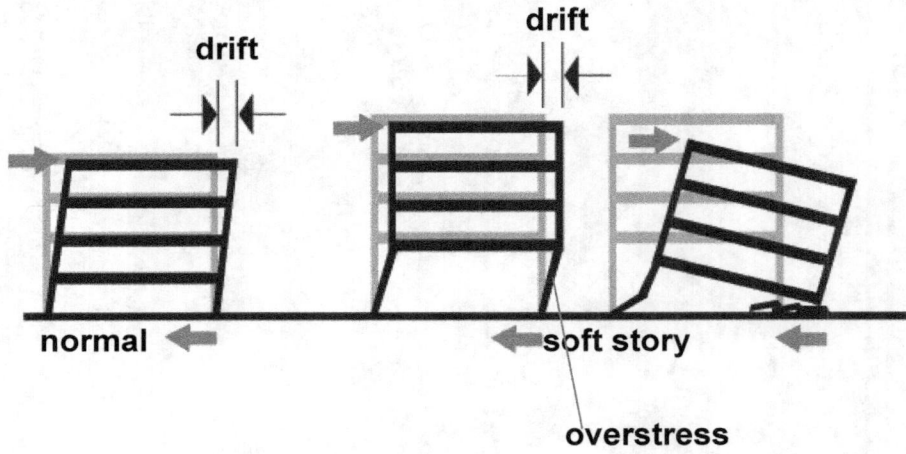

Figure 2-13 Deformation pattern due to soft-story behavior. Deformation is concentrated in the first story.

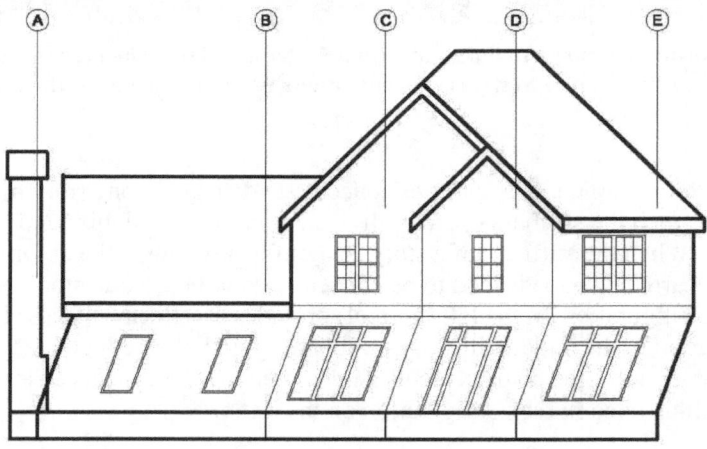

Figure 2-14 Change in deformation patterns associated with soft-story irregularity.

Soft and weak stories, although technically different, can occur at the same time. Combined soft- and weak-story house irregularities have been the primary cause of story failure or collapse and earthquake fatalities in wood light-frame houses in the United States. To date, story failure has only been observed in houses that would not meet the current *IRC* bracing requirements or would fall outside the scope of the *IRC*. Figure 2-15 illustrates soft- and weak-story behavior in a single-family house. Figure 2-16 illustrates the loss of a soft and weak story in an apartment building.

Figure 2-15 House experiencing soft- and weak-story behavior in the Cape Mendocino earthquake. Deformed shape of garage door opening shows approximately 6 inches of horizontal movement in the first story.

Figure 2-16 Apartment building experiencing soft- and weak-story behavior in the Northridge earthquake. The two balcony rails with little separation at the center of the photo illustrate the collapse of the lowest story.

IRC Approach to House Configuration Irregularities – The *IRC* incorporates two approaches to limitation of irregularities. The first and more explicit approach is found in *IRC* Section R301.2.2.2.2, which directly limits a series of irregular house configurations. Specific exceptions allow inclusion of less significant irregularities within *IRC* prescriptive designs. For one- and two-family detached houses, these provisions are applicable in SDCs D_1 and $D_{2,}$. The *IRC* limitations on irregularities are derived from those required for engineered buildings by the *IBC* and *NFPA 5000*. Table 2-3 at the end of this chapter provides detailed discussion of the *IRC* Section R301.2.2.2.2 limits. Additional discussion may be found in the commentary to the 2003 *NEHRP Recommended Provisions*.

The second and less obvious approach is the *IRC* requirement for distribution of wall bracing. Along with braced wall lines at exterior walls, interior braced wall lines must be added so that the spacing between wall lines does not exceed 35 feet per *IRC* Section R602.10.1.1 (adjustable to 50 feet by an exception) or 25 feet in SDC D_1 or D_2 per Section R602.10.11. Maximum spacing between braced walls in a braced wall line is also regulated. Houses perform better if the walls are distributed throughout, rather than concentrated in limited portions of the structure. This allows the earthquake load to be resisted local to the area where it is developed. Higher earthquake loads develop when loading must be transmitted to bracing walls in another portion of the house. Good distribution of bracing walls helps to mitigate the adverse effects of many irregularities.

Stepped Cripple Walls – IRC Section R602.11.3 and Figure R602.11.3 provide specific detailing to reduce the effect of a concentration of load due to stepped concrete or masonry foundation walls. As illustrated in Figure 2-17, a direct tie to the tallest foundation segment provides uniform stiffness along the wall line.

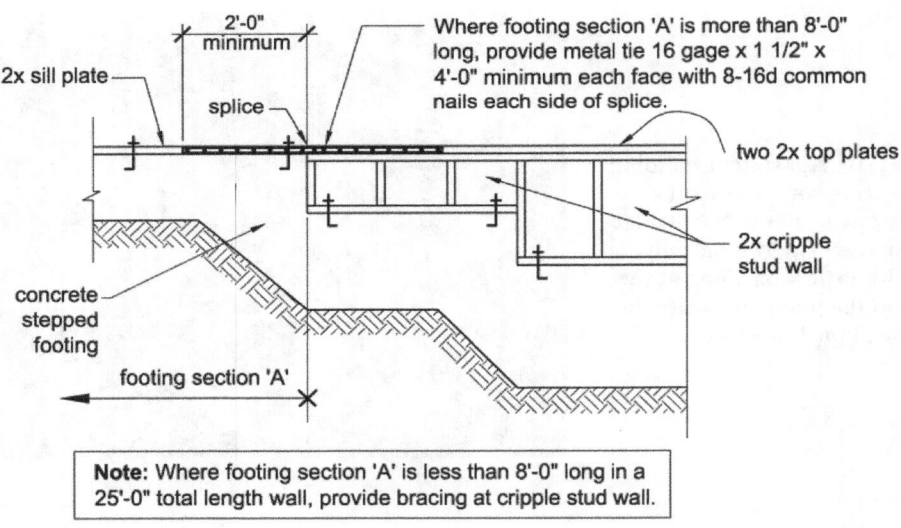

Figure 2-17 Detailing required for stepped foundation walls (ICC, 2003).

> **Above-code Recommendations:** The concentration of damage as a result of irregular building configuration is equally applicable to earthquake loading in all SDCs and to wind loading. To date, the *IRC* has limited application to areas of high seismic risk. **Use of the irregularity limitations in SDCs A, B, and C and for wind loading will contribute to better building performance and is recommended as an above-code measure.**
>
> Experimental and analytical results show that typical residential buildings are prone to soft-story irregularities. **Stiffening and strengthening of soft first stories is recommended as an above-code measure to reduce deformation and resulting damage.** Approaches to increasing first-story strength and stiffness include: increasing the length of wood structural panel sheathing; decreasing center-to-center spacing on sheathing nailing (although spacing of less than 4 inches is not recommended due to the possibility of splitting the framing and because other weak links could develop); using perforated shear wall construction (also called "continuously sheathed wall"); using segmented shear wall construction (also called "wood structural panel sheathed with overturning anchors"); or providing wood structural panel sheathing on both faces of bracing walls. See Chapter 6 of this guide for detailed descriptions of these measures.
>
> Another common location of soft or weak stories is cripple walls. This is particularly true where perimeter cripple walls have no inside face sheathing and under-floor basement areas have few or no interior walls. Although the *IRC* has increased bracing wall lengths beyond those required by earlier codes, further increasing length and providing interior braced wall lines will help to limit deformation and damage. **Additional length of wood structural panel sheathing at the building perimeter and interior is recommended as an above-code measure to limit cripple wall deformation.**

Table 2-1 Load Path Connections for Horizontal Sliding

Item	Minimum Fastening per *IRC* Table R602.3(1) and Discussion	Illustration
H1	**Sheathing**[a] **Nailing**[b] 5/16" to ½" 8d common @ 6" 19/32" to 1" 8d common @ 6" 1⅛" to 1¼" 10d common @ 6" • Resists roof sheathing sliding with respect to blocking below. • Six-inch nail spacing applies to supported sheathing edges and blocking. Twelve-inch spacing applies at other panel supports. • Rafter blocking is not always required by *IRC*; however, sheathing should be nailed to blocking where blocking is provided.	
H2	Three 8d box (0.113"x2½") or three 8d common (0.131x2½") toenails each block. • Resists rafter blocking sliding with respect to wall top plate. • **Use of angle clips in lieu of toenails is a recommended above-code measure.** • Rafter blocking is not always required by *IRC*; however, it should be fastened where provided.	
H3 & H4	**Sheathing**[a] **Nailing**[b] 5/16" to ½" 6d common @ 6" 19/32" to 1" 8d common @ 6" 1⅛" to 1¼" 10d common @ 6" • Provides wall racking resistance. • Six-inch nail spacing applies to sheathing edges. Twelve-inch spacing applies at other studs.	

Item	Minimum Fastening per *IRC* Table R602.3(1) and Discussion	Illustration
H5	**At Braced Wall Panels** Three 16d box (0.135"x3½") or three 16d sinker (0.148x3¼") face nails each 16 inches on center (space evenly). **Between Braced Wall Panels** One 16d box (0.135"x3½") or one 16d sinker (0.148x3¼") face nail at 16 inches on center. • Resists wall sole plate sliding with respect to sheathing and blocking or rim joist below.	
H6	Sheathing[a] Nailing[b] 5/16" to ½" 6d common @ 6" 19/32" to 1" 8d common @ 6" 1⅛" to 1¼" 10d common @ 6" • Resists floor sheathing sliding with respect to blocking below. • Six-inch nail spacing applies to supported sheathing edges and blocking. Twelve-inch spacing applies at other panel supports.	
H7	Three 8d box (0.113"x2½") or three 8d common (0.131x2½") toenails each block. • Resists joist blocking sliding with respect to wall top plate. • **Use of angle clips in lieu of toenails is a recommended above-code measure.**	

FEMA 232, Homebuilder's Guide

Item	Minimum Fastening (*IRC* Table R602.3(1) U.O.N.) & Discussion	Illustration
H8 & H9	**Sheathing**[a] **Nailing**[b] 5/16" to 1/2" 6d common @ 6" 19/32" to 1" 8d common @ 6" 1 1/8" to 1 1/4" 10d common @ 6" • Provides wall racking resistance. • Six-inch nail spacing applies to sheathing edges. Twelve-inch spacing applies at other studs.	
H10	Anchor bolts in accordance with *IRC* Sections R403.1.6 and R403.1.6.1. Steel plate washers in accordance with R602.11.1. Requirements vary by SDC. See Chapter 4 of this guide for further discussion. • Resists foundation sill plate sliding with respect to slab-on-grade or other foundation.	
H11	Foundation embedment in accordance with *IRC* Section 403.1.4 provides for development of lateral bearing and friction, which permits transfer of loads between the foundation and supporting soil. • Resists foundation sliding relative to soil (grade).	

a. Wood structural panel sheathing; see *IRC* Table R602.3(1) for other sheathing materials.
b. Common nail diameter and length: 6d: 0.113"x2", 8d: 0.131"x2-1/2", 10d: 0.148"x3".

Table 2-2 Load Path Connections for Overturning

Item	Overturning Load Path Description and Discussion	Illustration
OV1	When Wall 1 is loaded from left to right, the wall tries to overturn causing the lower left corner to uplift. This illustration shows a hold-down strap restraining this uplift. The hold-down strap carries tension from an end post or studs in the second-story wall to an end post or studs in the first-story wall, which in turn must be anchored to the foundation (OV5)	
OV2	When Wall 1 is loaded from left to right and an uplift load occurs at OV1, an approximately equal downward load occurs at OV2. This load will be in the post or studs at the end of the wall and will push down on the floor framing and first-story wall. This load will be transmitted through a first-story post to the foundation. When Wall 1 is loaded from right to left, there is an uplift load in the hold-down strap at OV2. Because this end of the wall is not aligned with a wall end in the first story, attention is needed to make sure that a post is added in the first-story wall for strap nailing. The first-story post can have edge nailing to the wall sheathing and transfer the uplift into the first-story wall or, alternatively, can be anchored directly to the foundation with an additional hold-down anchor. One or the other of these anchorage methods is needed to complete the load path.	
OV3	When Wall 2 is loaded from left to right, the wall tries to overturn causing the lower left corner to uplift. The location of OV3 over a first-story header makes the load path more complex than OV1. Hold-down strap OV3a carries the uplift load from the Wall 2 end post to the first-story header. Because the uplift load can be larger than the minimum load on the header, straps OV3b and OV3c are shown tying the header down to the first-story posts or studs. If this is not done, it might be possible for the header to pull up. When Wall 2 is loaded from right to left, a downward load occurs at OV3a. This downward load adds to the load already in the header and the studs supporting the header. When Wall 2 extends more than a foot over the header, the condition is considered an irregularity and is subject to limitations in SDC D_1 and D_2.	
OV4	When Wall 2 is loaded from left to right and an uplift load occurs at OV4. This load will be in the post or studs at the end of the wall and will push down on the floor framing and first-story wall. This load will be transmitted through a first-story post to the foundation.	

Item	Overturning Load Path Description and Discussion	Illustration
OV5	When Wall 3 is loaded from left to right, the wall tries to overturn causing the lower left corner to uplift. This illustration shows a hold-down strap restraining this uplift. The hold-down strap carries tension from an end post or studs in the first-story wall to the foundation. The uplift load in the first-story end post or studs is a combination of the second-story uplift load from OV1 and the uplift load accumulated over the height of the first story. Hold-downs anchored to the foundation should be used only where substantial continuous foundations are provided. Hold-downs anchored to existing foundations that are weak or that do not meet current dimensional requirements require engineering guidance. Hold-downs anchored to isolated footings require engineering guidance.	
OV6	When Wall 3 is loaded from left to right and an uplift load occurs at OV5, an approximately equal downward load occurs at OV6. This load will be in the post or studs at the end of the wall and will push down on the foundation. This load will be a combination of the downward load OV2 from Wall 1 and the load accumulated over the height of Wall 3. An exact engineering calculation would adjust this downward load based on the narrower width of Wall 2 and the uplift from the hold-down at OV3b.	
OV7	When Wall 4 is loaded from left to right, the wall tries to overturn causing the lower left corner to uplift. This illustration shows a hold-down strap restraining this uplift. The hold-down strap carries tension from an end post or studs in the first-story wall to the foundation. The uplift load in the first-story end post or studs is a combination of the second-story uplift load from OV3c and the uplift load accumulated over the height of the first story.	
OV8	When Wall 4 is loaded from left to right and an uplift load occurs at OV7, an approximately equal downward load occurs at OV8. This load will be in the post or studs at the end of the wall and will push down on the foundation. This load will be a combination of the downward load OV4 from Wall 2 and the load accumulated over the height of Wall 4.	

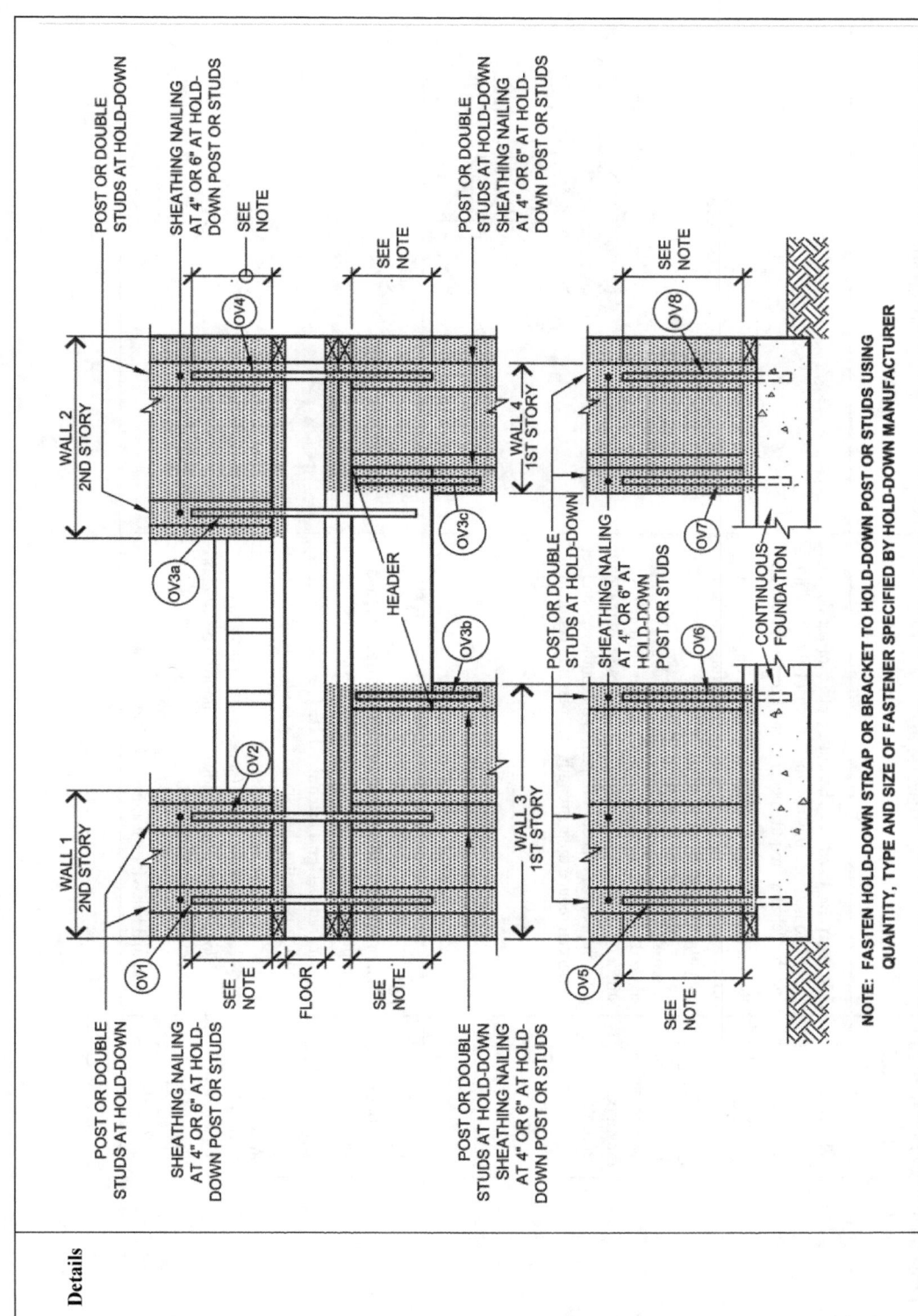

Table 2-3 *IRC* House Configuration Irregularities.

Irregularity Number	*IRC* Section R301.2.2.2 Description	Discussion	Illustration
1	When exterior shear wall lines or braced wall panels are not in one plane vertically from the foundation to the uppermost story in which they are required.	***IRC* May Be Used If:** The out of plane offset does not exceed four times the nominal depth of the floor joists and the detailing requirements of the exception are met. **Engineered Design Is Required If:** The out-of-plane offset exceeds that permitted or the detailing provisions are not followed. **Discussion:** Braced wall panels are intended to be stacked in order to maintain uniform strength and stiffness at each story and to aid in continuity of the load path. Support of braced wall panels on cantilevers and set backs can reduce their strength, stiffness and continuity.	
2	When a section of floor or roof is not laterally supported by shear walls or braced wall lines on all edges. Also called an "Open Front" irregularity.	***IRC* May Be Used If:** A section of floor or roof extends not more than six feet beyond the braced wall line, and the cantilever does not support a braced wall panel. **Engineered Design Is Required If:** The floor or roof extension is greater than six feet or is supporting a braced wall panel. **Discussion:** Placement of bracing walls at the edge of each floor or roof section aids in the uniform distribution of bracing and aids in ensuring continuity of the load path. Where bracing walls are not provided at all edges, rotational behavior (plan irregularity) may result.	

Irregularity Number	IRC Section R301.2.2.2 Description	Discussion	Illustration
3	When the end of a braced wall panel occurs over an opening in the wall below and ends at a horizontal distance greater than 1 foot from the edge of the opening. This provision is applicable to shear walls and braced wall panels offset out of plane as permitted by the exception to Item 1 above.	**IRC May Be Used If:** The braced wall panel does not extend more than one foot over the header below, or if the header meets the requirements of the exception. **Engineered Design is Required If:** The braced wall panel extends more than 1 foot over a header and the header is not selected in accordance with exception requirements, or if the entire braced wall panel falls on the clear span of the header. **Discussion:** When earthquake loads are applied to braced wall panels, large downward loads develop at the panel ends due to overturning (see Table 2). If the downward load falls on a header that is not strong or stiff enough, the effectiveness of the bracing is reduced and localized damage may occur.	EXTERIOR ELEVATION — REQUIRED BRACED WALL PANEL — MORE THAN 1'-0"
4	When an opening in a floor or roof exceeds the lesser of 12 feet or 50 percent of the least floor or roof dimension.	**IRC May Be Used If:** Floor and roof openings are kept to minimum size, such as standard stair openings. **Engineered Design Is Required If:** Stair openings are enlarged, such as to create entry foyers or two story great rooms, or accommodate large skylights. **Discussion:** Large floor and roof openings can affect the uniform distribution of earthquake loads to bracing walls, resulting in increased deformations and concentration of damage in the floor, roof and bracing wall systems.	PLAN VIEW — MORE THAN b/2 IS IRREGULAR

FEMA 232, Homebuilder's Guide

Irregularity Number	IRC Section R301.2.2.2 Description	Discussion	Illustration
5	When portions of a floor level are vertically offset. Also called "Split Level" irregularity.	*IRC* **May Be Used If:** Floor framing on either side of a common wall is close enough in elevation so that straps or other similar devices can provide a direct tension tie between framing members on each side of the wall. **Engineered Design Is Required If:** Floor framing on either side of a common wall cannot be directly tied together. **Discussion:** This irregularity results from observed earthquake damage in which one of two floor or roof levels pulled away from a common wall, resulting in local collapse. The direct tie limits the distance that either floor system can pull away, reducing likelihood of losing vertical support.	*(Illustration: Two diagrams showing common wall with offset in floor level. Top shows "STRAP FOR TENSION TIE", "COMMON WALL", "OFFSET IN FLOOR LEVEL" — labeled IRC MAY BE USED. Bottom shows "FLOOR OFFSET TOO LARGE FOR DIRECT TENSION TIE" — labeled ENGINEERED DESIGN REQUIRED.)*
6	When shear walls and braced wall lines do not occur in two perpendicular directions.	*IRC* **May Be Used If:** Required braced wall panels are oriented in the house longitudinal and transverse directions. **Engineered Design Is Required If:** Required bracing walls fall at angles other than longitudinal and transverse. **Discussion:** It has become somewhat common for houses to have walls that fall at an angle to the main transverse and longitudinal directions (often at 45 degrees). Where angled walls are not required for bracing, this is not a concern. Walls used for bracing must be aligned in the longitudinal or transverse direction. When the angle of bracing walls varies, the earthquake loads in the walls vary from those assumed in developing the *IRC* provisions. Non-standard load path detailing may also be required.	*(Illustration: Roof plan showing rectangular outline with an angled wall section labeled "ANGLED WALL NOT PERMITTED AS REQUIRED BRACING". Labeled ROOF PLAN.)*

Irregularity Number	IRC Section R301.2.2.2 Description	Discussion	Illustration
7	When stories above grade partially or completely braced by wood wall framing in accordance with Section R602 or steel wall framing in accordance with Section R603 include masonry or concrete construction.	**IRC May Be Used If:** Concrete or masonry construction within a light-frame house is limited to those items listed in the exception (fireplaces, chimneys and veneer). **Engineered Design Is Required If:** Other concrete or masonry construction is mixed with light-frame walls in any story above grade. **Discussion:** The *IRC* wall bracing requirements for wood or steel light-frame walls are proportioned to resist earthquake loads from light-frame wall systems only. Introduction of concrete or masonry will increase earthquake loads beyond the wall bracing capacity. In addition, introduction of concrete or masonry walls will likely effect the distribution of wall stiffness, causing a plan irregularity.	

Chapter 3
FOUNDATIONS AND FOUNDATION WALLS

This chapter discusses foundations and foundation walls constructed using the two most common foundation materials – concrete and masonry. Although the *IRC* permits the use of treated wood for foundations and foundation walls and insulating concrete form (ICF) for foundation walls, this guide does not cover those materials other than to inform the reader that wood foundations in Seismic Design Categories (SDCs) D_1 and D_2 require engineering design and many of the reinforcement requirements for concrete also apply to ICF (see Chapter 4 of the *IRC* for more information on the use of these materials). The subject of frost protection of foundations also is not discussed in this guide but, where required by the code (see *IRC* Section R403.1.4.1) or local regulations, foundations must either extend below the frost line or be protected from frost using approved methods.

3.1 GENERAL FOUNDATION REQUIREMENTS

Foundations are the interface between a house and the supporting soils. Many issues must be considered when selecting a foundation system including site topography, soils conditions, retaining requirements, loading from the house above, frost depth, and termite and decay exposure. Foundations primarily provide support for vertical gravity loads from the weight of a house and its contents, but they also provide resistance to horizontal sliding resulting from earthquake ground motions and must resist vertical loads at the ends of braced walls. Regardless of Seismic Design Category, all houses require a continuous foundation extending at least 12 inches below undisturbed soil along all exterior walls as shown in Figure 3-1.

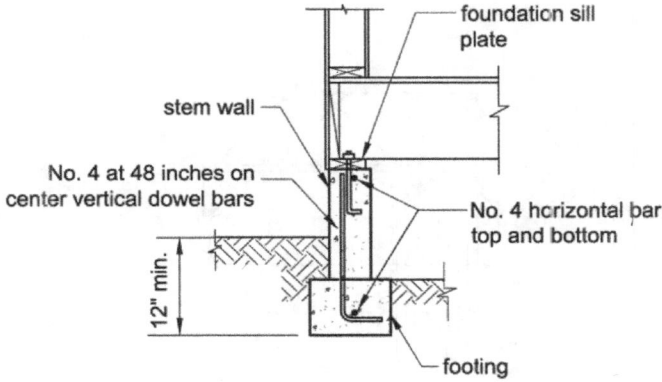

Figure 3-1 Perimeter foundation with separately placed footing and stem wall.

When earthquake ground motion occurs, the resulting ground movements, velocities, and accelerations are imparted to the foundation and, in turn, transferred to a house or other building. How well the house performs during an earthquake depends on the foundation being able to provide:

- Continued vertical support,

- Friction and passive bearing at the soil-to-foundation interface to minimize movement and damage,

- Anchorage at the foundation-to-house interface to minimize movement and damage, and

- Strength and stiffness sufficient to resist both horizontal loads and vertical loads resulting from racking and overturning of bracing walls within the house.

The foundation of the house must resist the sliding and overturning actions associated with an earthquake. These two actions are illustrated in Figures 3-2 and 3-3. The soil surrounding a foundation can resist sliding using a combination of friction along the bottom and bearing along the sides of the foundation; therefore, a wider and deeper foundation provides greater friction and greater bearing resistance than a shallow and narrow foundation. The whole overturning action illustrated in Figure 3-3 is resisted at the foundation in two ways. The portion of the foundation being pushed downward will bear against the soil below so a wider footing will provide more surface area to resist that downward load. At the uplift end of the foundation, the weight of any soil located above a footing helps to resist the loads trying to pull the foundation out of the ground; therefore, a deep inverted T-shaped foundation will provide greater resistance to uplift than a shallow footing or than a foundation having a shape that avoids having any soil above the top of the footing.

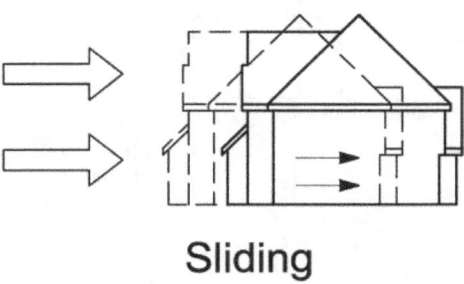

Figure 3-2 Sliding action resisted by foundation.

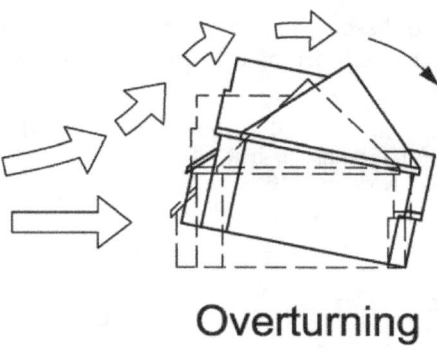

Figure 3-3 Overturning action resisted by foundation.

The *IRC* discusses foundations (footings and stem walls) and foundation walls separately and contains requirements for those elements based on the materials used for their construction. This guide chapter is similarly organized and addresses foundations placed monolithically with a slab on grade and foundations having a combination footing and stem wall as illustrated in Figure R403.1(1) of the *IRC*. Foundation systems such as pilings, drilled piers, and grade beams require the involvement of a licensed design professional and are not discussed in this guide.

IRC foundation wall provisions have evolved from similar provisions in the *CABO One- and Two-Family Dwelling Code* and *Standard Building Code*. Neither the *IRC* nor the other building codes give definitive guidance on when to use the foundation wall provisions of *IRC* Section R404 rather than the footing stem wall provisions of *IRC* Section R403. The 2000 *IRC Commentary* notes that the foundation wall provisions are primarily for masonry and concrete basement walls. The provisions of *IRC* Section R404 become mandatory for wall heights of 5 feet and greater and for walls retaining unbalance fill of 4 feet or greater. For wall heights and unbalanced fill heights less than this, there are few practical differences between the foundation and foundation wall provisions.

IRC Section R403.1.2 contains a general rule applicable to buildings located in SDCs D_1 and D_2 that requires interior braced wall lines to be supported on a continuous foundation when the spacing between parallel exterior wall lines exceeds 50 feet. However, *IRC* Section R602.10.9 contains a slightly more restrictive requirement. For a two-story house in SDC D_2, a continuous foundation is required below all interior braced walls, even when the distance between exterior walls does not exceed 50 feet, unless three additional conditions can be met. Those conditions are:

- The distance between braced wall lines does not exceed twice the building width measured parallel to the interior braced wall line;

- In houses having either a crawl space or basement, cripple walls cannot exceed 4 feet in height; and

- In houses having a crawl space or basement, first-floor interior braced walls are supported on double joists, beams, or blocking as shown in Figure 3-4.

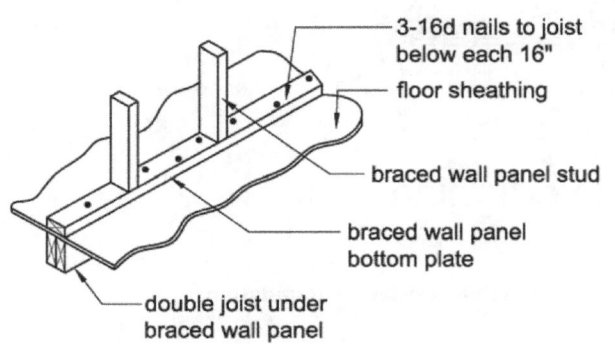

Figure 3-4 Interior braced wall on floor framing.

Above-code Recommendation: The main consequence of not providing a continuous foundation below a first-story interior braced wall line is that the floor must be strong enough to transfer earthquake loads from that interior location to a parallel perimeter foundation. This transfer of lateral loads can be accomplished using a wood-framed and -sheathed floor system, but this solution will definitely impart additional stresses in the floor and into cripple walls at the perimeter that would not occur if a foundation was provided below the interior braced wall line. Vertical loads due to the overturning loads in the braced wall segments also must be transferred to the perimeter foundation through bending action. **While the doubling of the floor joist is an improvement, supporting these walls directly on a continuous foundation is recommended to achieve above-code performance.**

For slab-on-grade construction in SDCs D_1 and D_2, when the conditions described above require an interior braced wall line to have a foundation, the foundation depth along that interior wall must be at least 18 inches below the top of the slab as shown in Figure 3-5.

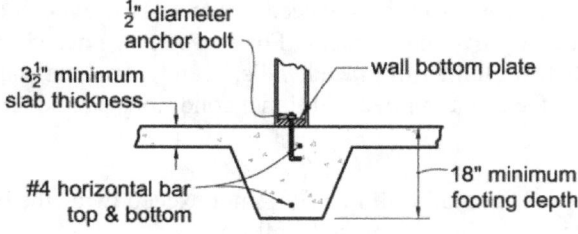

Figure 3-5 Foundation requirements for interior braced wall line on slab-on-grade construction.

3.2 CONCRETE FOUNDATIONS

Regardless of Seismic Design Category, the minimum specified concrete strength for foundations (and foundation walls) is 2,500 pounds per square inch (psi) with higher strength necessary when a foundation is exposed to the weather and the house is located in a moderate or severe weathering probability area as shown in *IRC* Figure R301.2(3). Specifying 2,500 psi refers to a measure of the concrete's compressive strength. To enable a concrete foundation to resist all of the possible loads to which it may be exposed, compressive strength needs to be complimented with tension capacity. Since concrete is unable to resist tension stresses without cracking, steel reinforcing bars are added to resist tension. Reinforcing is particularly valuable when resisting cyclic earthquake loads because, within the span of a few seconds, the loads may start by causing compression and then reverse to cause tension in the same location.

IRC Section R403.1.3 specifies minimum reinforcement of concrete footings located in SDCs D_1 and D_2. Separate subsections within *IRC* Section R403.1.3 address reinforcing of foundations consisting of a footing and a stem wall and reinforcing of the footing along the perimeter of a slab-on-grade. *IRC* Section R403.1.3 also contains an exception that allows omitting the longitudinal reinforcing in concrete footings for houses that are three stories or less in height and constructed with stud walls, regardless of the Seismic Design Category.

> **Above-code Recommendation: To obtain above-code performance in SDC C, it is recommended that the minimum foundation reinforcing requirements for SDCs D_1 and D_2 be used.** This added reinforcing will provide better footing performance whether it is resisting earthquake loads or loads induced by differential soil settlement, expansive soils, or frost heave.

Typically, the bottom portion of a concrete footing must have one horizontal No. 4 reinforcing bar located 3 inches up from the bottom of the concrete (clear from the soil along the bottom of the footing). When a foundation consists of both a footing (horizontal foundation segment) and short stem wall (vertical foundation segment), two No. 4 continuous horizontal reinforcing bars are required – one in the bottom of the footing and one near the top of the stem wall as shown in Figure 3-6.

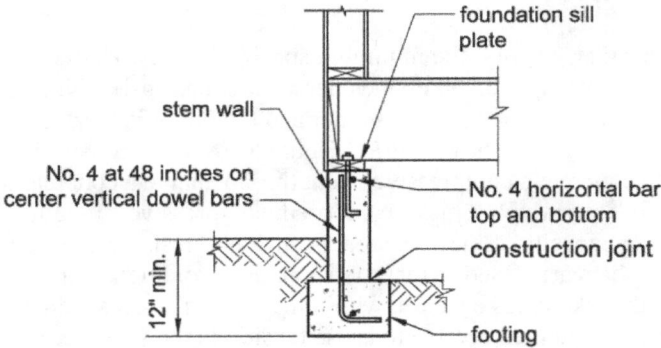

Figure 3-6 Recommended minimum reinforcement for concrete footings and stem walls.

Above-code Recommendations: When horizontal reinforcing bars are used in foundations, they need to be continuous to perform their intended function. The *IRC* does not provide any specific guidance on reinforcing continuity; therefore, the following **above-code** recommendation is derived from the basic standard for concrete construction (ACI 318-02, 2003). **Where two or more pieces of reinforcing steel are used to provide continuous horizontal reinforcing, the ends of the bars should be lapped to provide continuity. The minimum recommended lap for No. 4 bars is 24 inches and for No. 5 bars is 30 inches. As shown in Figure 3-7, horizontal bars terminating at corners of perimeter foundations and where an interior foundation intersects a perimeter foundation should have a standard 90-degree hook of 8 inches for No. 4 bars or 10 inches for No. 5 bars.**

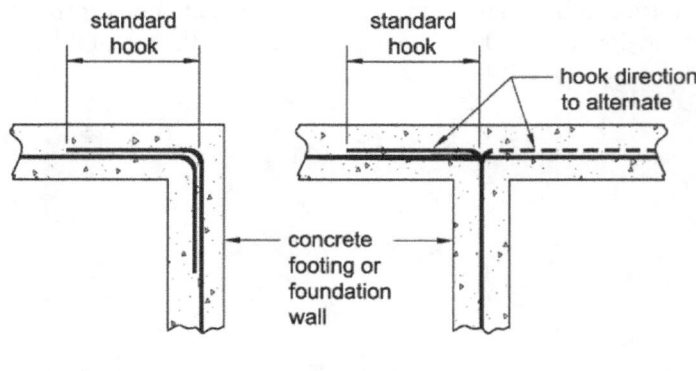

Figure 3-7 Above-code horizontal reinforcing lap at corners and intersections.

In SDCs D_1 and D_2, No. 4 vertical reinforcing is required at 48-inch maximum spacing if a construction joint (also called a "cold joint") occurs between the footing and the stem wall. This often is referred to as a "two-pour" foundation because the concrete for the footing and the concrete for the wall are poured at separate times. These vertical bars (also called "dowels") must extend a minimum of 14 inches into the stem wall and be hooked at the bottom into the footing as shown in Figure 3-6. These dowels provide a very important connection because, without them, earthquake loads can cause sliding to occur along the joint between the two separate concrete placements. Sliding along a similar construction joint between a slab-on-grade and a separately poured footing below the slab edge occurred during the 1994 Northridge earthquake and caused severe damage in houses located in Simi Valley, California.

When a house has a concrete slab-on-grade with a thickened edge forming its perimeter foundation (also called a "turned-down slab edge"), one No. 4 horizontal reinforcing bar is required in the top and bottom of this footing as shown in Figure 3-8. The exception to this is when the slab and footing are poured at the same time; in this case, a single No. 5 bar or two No. 4 bars located in the middle third of the combined slab and footing depth may be used. For slab-on-grade construction in SDCs D_1 and D_2, interior bearing walls and interior braced walls required to have a continuous foundation must have the concrete slab thickened to 18 inches to form a foundation as shown in Figure 3-5.

> **Above-code Recommendation:** Although the *IRC* is basically silent on how to reinforce these interior thickened slab foundations, it is recommended that they should be reinforced as described above for the foundation along the slab perimeter. Horizontal reinforcing in slab-on-grade foundations should be continuous as described earlier.

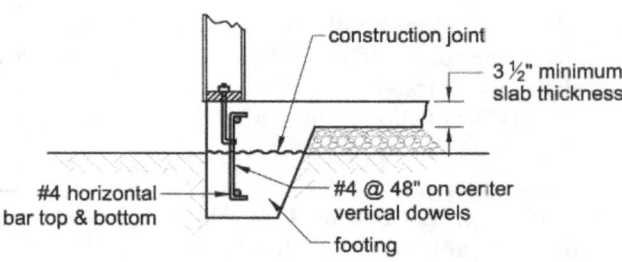

Figure 3-8 Above-code use of vertical dowels to connect a slab-on-grade to a separately poured footing.

> **Above-code Recommendation:** When a slab-on-grade is poured separately from the footing below, the *IRC* does not specify any vertical reinforcing across this joint. When this condition occurs, however, the possibility of earthquake loads causing sliding along that joint is very real. The basic provisions of ACI 318-02, Sec. 11.7.9, for concrete require that all construction joints be provided with a mechanism for transferring loads through the joint. **Therefore, it is recommended that the same amount of vertical reinforcing required for a separately poured stem wall and footing condition (No. 4 at 48 inches on center) be installed to connect a slab to a separately poured footing as shown in Figure 3-8. In addition the surface of the construction joint should be cleaned to remove loose debris prior to placing the concrete slab.**

> **Above-code Recommendation:** Anchor bolts connecting a wood sill plate to a foundation also must be installed in the foundation or stem wall. **Although it is possible to install anchor bolts during the placement of the foundation concrete before it has hardened (also called "wet setting"), this is not recommended. Anchor bolts should be installed and secured so that they will not move prior to placing the concrete.** The reason is that a wet setting can create a void in the concrete adjacent to the bolt. Anchor bolts, whether they have heads or hooks on the embedded end, will either make a hole or carve a slot as they are pushed into the wet concrete. That hole or slot forms a void over the entire length of the bolt's embedment. This void prevents the bolt from completely bearing against the surrounding concrete and could result in movement of the bolt when subjected to earthquake loads. Additional anchor bolt requirements are provided at the end of this chapter.

3.3 MASONRY FOUNDATIONS

Masonry foundation requirements are generally more dependent upon the Seismic Design Category of the site than concrete foundation requirements. Solid clay masonry or fully grouted concrete masonry and rubble stone masonry may be used for foundations in SDCs A, B, and C; however, rubble stone masonry is not allowed in SDCs D_1 and D_2 due to its relatively low strength and stability.

> **Above-code Recommendation:** Because of their relatively low strength and stability, **rubble stone foundations are not recommended for use in SDC C as an above-code measure.**

> **Above-code Recommendation:** Like concrete, a masonry foundation with a stem wall must have at least one No. 4 horizontal bar in the bottom of the footing and one No. 4 bar near the top of the stem wall. This horizontal reinforcing needs to be continuous just as it does in concrete construction. **Therefore, the above-code recommendations for lap splices are 24 inches for No. 4 bars and 30 inches for No. 5 bars with hook extensions at corners as shown in Figure 3-7.**

In SDCs D_1 and D_2, masonry stem walls also must have vertical reinforcing and must be grouted as required by *IRC* Sections 606.11.3 and 606.12. (See Chapter 6 of this guide for a more detailed discussion of grouting in masonry wall construction.) For a masonry stem wall supported on a concrete footing, the vertical reinforcing must be one No. 4 bar at 48-inches maximum spacing, extending into the footing with a standard hook at the bottom, similar to what is shown in Figure 3-6. This is necessary to prevent sliding along the bottom mortar joint between the concrete footing and the masonry stem wall.

> **Above-code Recommendation:** In order to prevent sliding along the bottom mortar joint between the concrete footing and the masonry stem wall, it is recommended that masonry stem walls in SDC C also use the vertical reinforcing required in SDCs D_1 and D_2.

3.4 FOOTING WIDTH

Footing width is not dependent upon Seismic Design Category but instead is solely based on vertical load considerations. The minimum width for a concrete or masonry footing is dependent on the load bearing capacity of the soil measured in pounds per square foot (psf) and the number of stories and the weight of the wall it supports. For instance, when brick veneer is installed, this added weight requires a wider footing.

The minimum soil bearing capacity considered by the *IRC* is 1,500 psf. In this case, the minimum footing width is 12 inches but it can increase to as much as 32 inches for a three-story house with brick veneer as shown in Figure 3-9.

Bearing capacity is determined based on the soil classification determined for the site. Soil classifications and corresponding bearing capacities are listed in *IRC* Table R401.4.1 at the beginning of *IRC* Chapter 4. The soil classification system is described in *IRC* Table R405.1. There are four distinct groups of soils that comprise a total of 15 separate soil classifications ranging from well graded gravels as the best and peat as the worst. Accurate determination of soil bearing capacity requires correct classification of the soil at the building site and may require the expertise of a soils engineer or geologist. However, most building departments have determined the soil classification for most of the sites within their jurisdiction. Accurate determination of the correct site soil classification is important not just for determining the minimum footing width but also for determining the minimum reinforcing required for concrete or masonry foundation walls.

When a footing is constructed with a width at the bottom that is greater than the thickness of its stem wall or a width greater than the thickness required for a foundation wall (e.g., an inverted-T shape), the minimum thickness of the footing is 6 inches. *IRC* Section R403.1.1 also specifies both a minimum and maximum width for the footing portion of this inverted-T foundation. As a minimum, the projection of the vertical face of the footing must be at least 2 inches beyond each vertical face of the stem wall or foundation wall. The maximum projection may not exceed the thickness of the footing. These dimensional requirements for an inverted-T foundation are illustrated in Figure 3-9. These minimum and maximum projection limits of the footing beyond the stem wall are consistent with ACI 318 design of plain concrete footings. Based on those requirements, an L-shaped footing should not be permitted unless its dimensions and reinforcing are designed to account for the eccentricity of the vertical load on the footing.

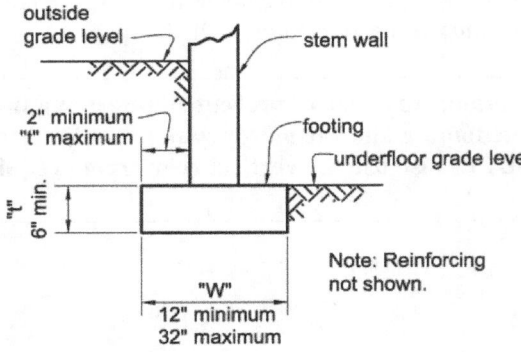

Figure 3-9 Inverted -T footing dimensions.

3.5 SPECIAL SOIL CONDITIONS

On some sites, such as those in marshy areas or bogs, the bearing capacity of the soil may be known or suspected to be less than 1,500 psf. In such a case, the *IRC* requires a soil investigation and report to determine the actual bearing capacity of the soil and to define an appropriate footing width and depth. Soil testing also is necessary when it is likely that existing soil has shifting, expansive, compressive, or other unknown characteristics. Local officials often are aware of such conditions in locations where previous buildings have been constructed and may have maps identifying such areas and/or special rules that apply to foundation construction in such areas. When existing soil data are not available, such as on previously undeveloped sites, it is always prudent to obtain tests to validate bearing capacity and to determine if expansive soils are present.

When expansive soils that exhibit large changes in volume (usually in response to changes in moisture content) are encountered, foundations must be designed in accordance with *IBC*

Chapter 18 because the prescriptive requirements of the *IRC* assume no such special conditions are present. Methods to address expansive soil include foundation designs to resist the stresses caused by the soil volume changes that are likely to occur, isolation of the foundation, removal of the expansive soil, or stabilization of the soil by chemical, dewatering, presaturation, or other methods. Failure to identify and adequately compensate for differential movements caused by expansive soils can result in excessive stresses on the foundation causing cracking of even reinforced concrete or masonry foundations. Foundation movement induced by expansive soil also can result in differential movement of the house's wood framed walls that can crack brick, gypsum wallboard, and stucco finishes. Movement of the walls can create stresses that loosen the nailed connections of wall sheathing used to provide lateral bracing. These effects can, in turn, weaken a building's earthquake resistance so it is very important to address expansive or other special soil conditions to limit differential foundation movement.

3.6 FOUNDATION RESISTANCE TO SLIDING FROM LATERAL LOADS

Foundations are the final link in the load path within a building to transfer the earthquake loads to the ground. At the foundation level, the combined lateral loads from the entire building are attempting to push the building laterally. To resist movement, the foundation pushes against the soil that surrounds it. Consequently, footing width and depth are factors that determine the resistance that can be provided by the foundation to earthquake ground motions. This is because footing width determines the horizontal surface area of the bottom of a foundation in contact with the ground and the depth determines the vertical surface area bearing against the soil on either side. These surface areas provide resistance to sliding through a combination of friction and bearing against the soil as shown in Figure 3-10. Therefore, a wider or deeper foundation will be capable of resisting greater lateral loads than a narrower or shallower foundation. Similarly, the sliding resistance of a slab-on-grade house will be greater than that of one having only a perimeter foundation due to the added frictional surface area provided by the underside of the slab.

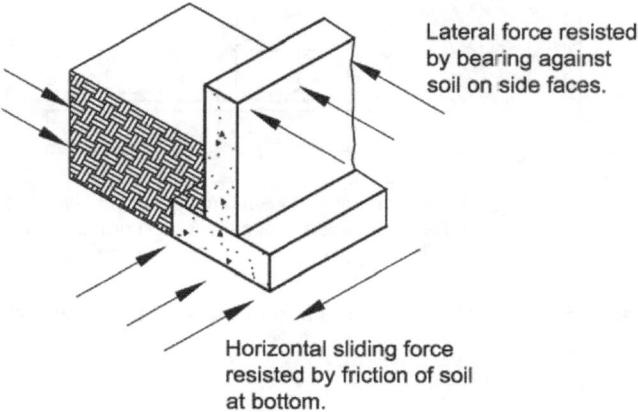

Figure 3-10 Lateral resistance provided by foundation.

FEMA 232, Homebuilders' Guide

> **Above-code Recommendation:** For houses with the first floor located above a crawl space rather than a basement, *IRC* Section 408.5 allows finished grade under the floor to be located at the bottom of the footing except where the groundwater table is high, surface drainage is poor, or the area is prone to flooding. However, when the finished grade is located at the bottom of the footing, the vertical face of the footing on that side does not bear against soil. **Therefore, as an above-code measure, in SDCs D_1 and D_2, it is recommended that crawl space perimeter footings have their entire depth below the finished grade of the crawl space.** Embedment along both vertical faces of the footing provides additional bearing surface area to resist sliding loads perpendicular to the exterior walls.

> **Above-code Recommendation:** To aid in providing sliding resistance, the bottom of footings should be level. The *IRC* does allow the bottom of footings to be sloped at a rate of not more than 1 foot vertical in 10 feet horizontal. When this maximum slope along the bottom of the footing cannot be met, the bottom must be stepped. Similarly, the top of foundations must be level but can also be stepped. **As an above-code measure when steps are used, the horizontal reinforcing should be bent to extend through the steps as shown in Figure 3-11.**

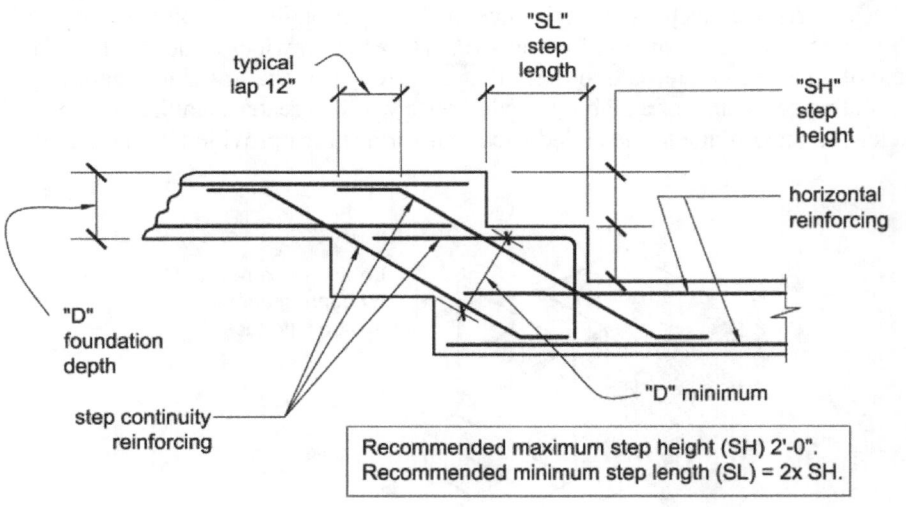

Figure 3-11 Above-code stepped foundation reinforcing detail.

60

Chapter 3, Foundations and Foundation Walls

> **Above-code Recommendation:** Special consideration also must be given to sites where the ground slopes upward or downward beyond the limits of the perimeter foundation to ensure that adequate resistance to foundation sliding and settlement are provided and to protect against the effects of drainage, erosion, and shallow failures of the sloping surface. Prescriptive rules for setbacks from either an ascending or descending slope are given in *IRC* Chapter 4. Smaller (or greater) setbacks may be approved (or required) based on a soil investigation report prepared by a qualified engineer. **Because landslides can be triggered by earthquakes, any building site having natural or man-made sloping terrain above or below should be thoroughly evaluated for landslide potential even when the prescriptive setbacks of the code are met.**

3.7 SPECIAL CONSIDERATIONS FOR CUT AND FILL SITES

A hillside site can result in foundations being supported on soils having very different bearing capacities. Figure 3-12 shows a situation where a portion of the foundation is supported on rock and the other side is supported on a fill that extends above the existing rock grade. This condition often occurs when soil is removed (cut) from the high side of a lot and the lower side has fill material installed to create a level building pad. Although the *IRC* requires all fill soils to be designed, installed and tested in accordance with accepted engineering practice, there is no specific guidance given in the *IRC* regarding what should be addressed. Site-specific guidance on the design and placement of fill material is a particularly important concern in high seismic areas.

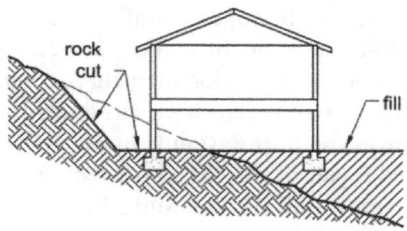

Figure 3-12 Foundation supported on rock and fill.

Studies of damage to houses located on cut and fill sites as a result of the 1971 San Fernando (McClure, 1973, Slosson, 1975) and the 1994 Northridge (Stewart, et al., 1994 and 1995) earthquakes document the consequences of improperly installed and compacted fill. Figure 3-13 shows an example of the type of damage that occurred to slab-on-grade houses during the Northridge earthquake. Generally, these sites experienced settlement and extension of the fill portion of the site and vertical displacement along the line of transition from cut to fill.

Above-code Recommendation: To meet the intent of the *IRC* requirement for all fill soils to be designed, installed and tested in accordance with accepted engineering practice, it is recommended that the reporting requirements conform to those specified in *IBC* Section 1803.5.

Figure 3-13 Example of damage caused in building on cut and fill site.

Above-code Recommendation: Even with proper installation and compaction of engineered fills, earthquakes are expected to result in some differential settlement and consequent damage when a house has a portion of its foundation on a cut pad and other portions on fill. Therefore, in SDCs C, D_1 and D_2, sites that require a cut and fill approach should be avoided, particularly if a slab-on-grade foundation is used. Where cut and fill grading of a site is unavoidable, increased levels of quality control should be used to ensure the optimum installation of the fill, and foundations should be designed to either accommodate or resist the expected settlement.

3.8 FOUNDATION WALLS

Foundation walls are typically basement walls but also include foundation stem walls that extend from the top of a footing to the bottom of a wood framed floor and enclose a crawl space as shown in Figure 3-6. Depending on the difference in ground level on each side of a foundation wall, certain minimum requirements for reinforcing and wall thickness will apply. Foundation wall construction is dependent to some extent on Seismic Design Category, particularly for SDCs D_1 and D_2 sites.

Foundation walls can be constructed of concrete, masonry, or preservatively treated wood or by using insulating concrete form (ICF) systems. As mentioned earlier, treated wood systems are not discussed in this guide. ICF foundation walls are likewise not discussed because *IRC* Section R404.4.1 limits use of ICF foundation walls to SDCs A, B and C; however, future *IRC* editions may allow adequately reinforced ICF foundation walls in SDCs D_1 and D_2 if proposed code changes on this subject are approved. Rubble stone masonry foundation walls also are not permitted in SDCs D_1 and D_2 for reasons similar to those that prohibit this material's use as a footing in SDC D_1 and D_2.

Regardless of Seismic Design Category, all foundation walls must extend a minimum of 6 inches above the adjacent grade or, when brick veneer is used on a wall, the foundation wall must extend a minimum of 4 inches above the adjacent grade.

Concrete and masonry foundation walls must conform to the prescriptive requirements of the *IRC* or may be based on other recognized structural standards such as ACI 318 for concrete or either NCMA TR68-A or ACI 530/ASCE 5/TMS 402 for masonry. When concrete or masonry walls are subject to hydrostatic pressure from groundwater or support more than 48 inches of unbalanced fill, without permanent lateral support at the top and bottom, they must be designed using accepted engineering practice. The discussion below relates only to the *IRC* prescriptive methods for concrete or masonry foundation walls.

Several terms used in this discussion warrant definition. "Plain concrete" and "plain masonry" are not necessarily devoid of all reinforcing; however, they have less reinforcing than is required to be officially designated as being "reinforced." "Unbalanced backfill" is defined as the difference in height between the exterior and interior finish ground levels except that, when an interior concrete slab floor is present, the height is measured from the top of the slab to the exterior finished ground level.

3.9 FOUNDATION WALL THICKNESS, HEIGHT, AND REQUIRED REINFORCING

The minimum thickness of plain concrete and plain clay or concrete masonry foundation walls ranges from 6 inches to 12 inches depending on several variables. These variables include:

- The height of the wall,
- The site's soil classification, and
- The height of any unbalanced backfill.

Foundation wall thickness also must be at least equal to the width of the supported wall.

For plain masonry foundation walls, the minimum thickness also is dependent on the use of solid masonry units or hollow units that can be either grouted or ungrouted. Generally, the minimum thicknesses for solid masonry and grouted hollow masonry are identical whereas the minimum thickness for ungrouted hollow masonry is greater. Minimum wall thickness for plain concrete and plain masonry walls, based on specific combinations of these variables, are listed in *IRC* Table R404.1.1(1).

For SDCs D_1 and D_2, *IRC* Section R404.1.4 imposes the following additional limitations on plain concrete and plain masonry foundation walls that are not specified in *IRC* Table R404.1.1(1):

- Wall height is limited to 8 feet.
- Unbalanced fill height is limited to 4 feet.
- A single horizontal No. 4 reinforcing bar shall occur in the upper 12 inches of the wall.
- Plain masonry walls shall be a minimum of 8 inches thick.
- Plain concrete walls shall be a minimum of 7.5 inches thick except that a 6-inch minimum thickness is permitted when the wall height does not exceed 4 feet 6 inches.
- Vertical reinforcing of masonry stem walls shall be tied to horizontal reinforcement located in the footing.

In SDCs D_1 and D_2, when the foundation exceeds 8 feet in height or supports more than 4 feet of unbalanced fill, additional minimum reinforcing requirements apply. First, two No. 4 horizontal reinforcing bars are required at the top of the wall rather than just one. In addition, a concrete or masonry wall's vertical reinforcing is required to meet additional prescriptive minimums. The vertical reinforcing size and maximum spacing are dependent on wall thickness, wall height, unbalanced fill height, and soil classification. Three *IRC* Tables — Tables R404.1.1(2), (3) and (4) – specify the minimum vertical reinforcing based on specific combinations of these variables. The *IRC* also allows alternative sizes and spacing of reinforcing up to a maximum spacing of 6 feet; however, the alternate size and spacing must result in an equivalent cross-sectional area of reinforcing per lineal foot of wall as prescribed in the tables.

> **Above-code Recommendation:** As described earlier, reinforcing should always be continuous; therefore, lap splices are needed where discontinuities occur in either horizontal or vertical reinforcing. Lap lengths should not be less than 24 inches for No. 4 bars or 30 inches for No. 5 bars.

Regardless of Seismic Design Category, all vertical reinforcing for these reinforced masonry or concrete foundation walls must be at least ASTM Grade 60 (yield strength of 60,000 psi). This is important to note because No. 4 bars are commonly available in Grade 40, which has a lower yield strength. In addition, the distance from the soil side face of the wall to the centerline of the vertical reinforcing must be 5 inches in an 8-inch-thick wall, 6.75 inches in a 10-inch-thick wall, and 8.5 inches in a 12-inch-thick wall.

3.10 WOOD-FRAMED WALL BOTTOM PLATE AND FOUNDATION SILL PLATE ANCHORAGE

For the purpose of the following discussion, the terms "wall bottom plate" and "foundation sill plate" are used to distinguish between two different locations of wood members along a foundation. A wall bottom (sole) plate either is directly supported by a slab-on-grade foundation as shown in Figure 3-5 or is part of a cripple wall supported directly on a foundation when a slab on grade does not occur. A foundation sill plate is different in that it occurs without wall studs framing on top of it; instead, a foundation sill plate is the bearing support between floor joists or floor beams and the foundation as shown in Figure 3-6.

In all Seismic Design Categories, where anchor bolts are required to connect a bottom plate or foundation sill plate to a foundation, the bolts must have a minimum embedment of 7 inches into a concrete or masonry foundation, and they must have a nut and washer tightened on each bolt. Typically, anchor bolts must be a minimum of 1/2-inch diameter and be spaced not to exceed 6 feet on center. However, in SDCs D_1 and D_2, the maximum bolt spacing is limited to 4 feet for buildings with more than two stories. Because the *IRC* prescribes bracing only for houses in SDC D_2 up to a maximum of two stories, the closer spacing is actually only applicable in SDC D_1 where three stories is permitted. This decreased bolt spacing provides 50 percent more shear capacity at the sill-plate connection in recognition that earthquake loads generated in a three-story building will be greater than those in a one- or two-story building.

Some builders may choose to use 5/8-inch diameter bolts at the standard 6 feet on center spacing in place of using 1/2-inch diameter bolts at a required 4 feet on center spacing. Based on bolt capacities listed in 2001 NDS Table 11-E, for most species of wood, 5/8-inch bolts at 6 feet on center spacing provide approximately 95 percent of the capacity of 1/2-inch bolts at 4 feet on center spacing. However, the approval of local officials should always be obtained prior to making any such substitution.

In SDCs D_1 and D_2, the washer to be installed under the anchor bolt nut must be a 3-inch by 3-inch square plate with a minimum thickness of 1/4 inch. This increased washer size limits the potential for splitting of bottom plates.

> **Above-code Recommendation:** Splitting of sill plates can occur when the ends of braced wall segments are lifted vertically during the rocking motion braced walls undergo while resisting earthquake loads. Cyclic testing of shear walls has shown that a standard round cut washer is too small to provide any appreciable resistance to this uplift and allows sill plates to split but, when larger square washers are used, walls can sustain much higher lateral loads before sill plate splitting will occur. **For this reason, the use of plate washers is also recommended on anchor bolts connecting to the foundation in SDC C.**

3.11 REQUIRED LOCATIONS FOR ANCHOR BOLTS ALONG EXTERIOR WALLS

Independent of Seismic Design Category, anchor bolts are required to connect a wall bottom (sole) plate or foundation sill plate to foundations located along all exterior wall lines. A minimum of two anchor bolts are required in each individual length of plate, with one bolt located not more than 12 inches nor less than 7 bolt diameters from each end.

3.12 REQUIRED ANCHORAGE ALONG INTERIOR BRACED WALLS

The anchorage of wall bottom plates along interior wall lines is a bit more complex than it is for exterior walls. The general rule is that when interior braced walls are supported directly on a continuous foundation, the bottom plate of the wall must be anchored using the same bolting pattern as required for exterior walls. However, the *IRC* currently requires a continuous foundation along interior braced wall lines in only a few situations. Consequently, bolting of the sole plate of an interior braced wall to a foundation occurs only rarely because a foundation is not normally required.

When an interior braced wall frames onto a raised wood framed floor, the code is clear that the anchorage connection of the braced wall sole plate uses nails. However, the *IRC* is silent regarding exactly what kind of anchorage is required for connecting an interior braced wall to a slab-on-grade when a continuous foundation is not required and not provided. One common example is a slab-on-grade house that has its exterior walls less than 50 feet apart and therefore does not require its interior braced walls be supported on a continuous foundation. In such a case, the slab usually will not have the thickness necessary to allow installation of anchor bolts with the standard 7 inches of embedment. However, it is reasonable to assume that the braced wall bottom-plate-to-slab connection should provide lateral load resistance at least equivalent to the typical anchorage to a foundation using bolts. For the 1/2-inch diameter bolts at 6 feet on center, the bolt shear capacity from NDS Table 11E, adjusted by 1.6 for earthquake loads, ranges from 152 plf in Spruce-Pine-Fir lumber, to 176 plf in Southern Pine lumber. Therefore, the connection to be provided to the slab should provide at least the same capacity as the 1/2-inch anchor bolts. For 5/8-inch bolts at 6 feet on center, the capacity ranges from 219 plf in Spruce-Pine Fir to 248 plf in Southern Pine.

When anchor bolts cannot be used because of limited slab-on-grade thickness, interior wall bottom plates often are connected to a slab-on-grade with powder-driven nails or pins as shown in Figure 3-14. However, because of their small diameter and shallow embedment length into the slab, each powder-driven nail or pin has significantly less lateral capacity than a single 1/2-inch diameter bolt. The nails or pins would need to be spaced at a much closer center-to-center spacing to be equivalent to a 1/2-inch diameter bolt at 6 feet on center. The actual spacing needed to achieve equivalence with anchor bolts generally depends on the diameter and length of the nail or pin.

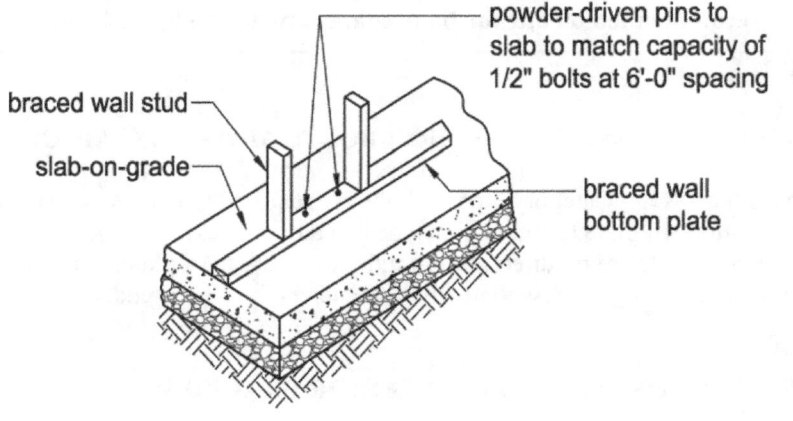

Figure 3-14 Interior braced wall on slab-on-grade.

Many manufacturers of powder-driven nails or pins have published shear and tension capacity values for their specific products when connecting wood members to concrete. However, it should be noted that powder-driven nails or pins used for this purpose generally have very little embedment depth into the concrete and, as a result, have very little tension capacity to resist the uplift that can occur at the ends of braced walls. Because of this limited tension capacity, they may not perform nearly as well as bolts. In addition, when pins are used, there is no way to install the square plate washers required on anchor bolts in SDCs D_1 and D_2 and recommended for use in SDC C.

> **Above-code Recommendation:** In SDCs C, D_1 and D_2, when interior braced walls use wood structural panel sheathing, it is recommended that bottom plates be anchored to a thickened slab-on-grade using bolts as shown in Figure 3-5. Instead of the 18-inch thickness shown in Figure 3-5, only a minimum 10-inch thickness is necessary for the slab-on-grade to allow for 7-inch minimum embedment of the bolt and 3 inches of additional clearance from the end of the bolt to the bottom of the thickened slab.

3.13 ANCHORING OF INTERIOR BRACED WALLS IN SDCs D_1 AND D_2

In SDCs D_1 and D_2, anchor bolts are required for interior first-story walls when:

- The first-story interior wall is a bearing wall (which may or may not be a braced wall line) and it is supported on a continuous foundation.

- The interior first story-wall is a braced wall line and that wall line is <u>required</u> to be supported on a continuous foundation.

When applying the two rules listed above, one must differentiate between interior walls that are bearing walls and those that are braced walls. Although interior bearing walls can be braced walls, this is not always the case. Therefore, to apply the rules it must be determined that:

- An interior wall is only a bearing wall and *not* a braced wall,
- An interior wall is a braced wall but is *not* a bearing wall, or
- The interior wall is *both* a bearing wall and braced wall.

Summaries of the 2003 *IRC* minimum requirements for continuous foundations and for installing anchor bolts to a foundation along braced wall lines and bearing walls are presented in Tables 3-1 and 3-2, respectively. When an interior wall is both a bearing wall and a braced wall, the most restrictive requirement from Tables 3-1 and 3-2 applies.

Table 3-1 Summary of 2003 *IRC* Continuous Foundation and Anchor Bolt Requirements for Braced Wall Lines in One- and Two-family Houses

Seismic Design Category	Continuous Foundation Required At Braced Wall Lines	Anchor Bolts Required	Anchor Bolt Diameter Spacing
A, B and C	YES along Exterior Braced Walls	YES	½" @ 6'-0"
A, B and C	NO along Interior Braced Walls	NO	N/A
D_1	YES along Exterior Braced Walls	YES[1]	½" @ 6'-0" ½" @ 4'-0" >2 story
D_1	NO along Interior Braced Walls Unless wall lines spaced > 50 feet	NO	N/A
D_2	YES along Exterior Braced Walls	YES[1]	½" @ 6'-0" [3]
D_2	NO along Interior Braced Walls of 1 Story unless spaced > 50 feet	NO	N/A
D_2	YES along Interior Braced Walls of Two-Story[2]	YES[1,2]	½" @ 6'-0" [3]

[1] Requires a square plate washer 3 x 3 x ¼ inch on each bolt.
[2] A continuous foundation is NOT required for interior braced wall lines of a two story building in SDC-D_2, provided that all of the following conditions are met: A) The spacing between continuous foundations does not exceed 50 feet, B) cripple walls (if provided) do not exceed 4 feet in height, C) first story braced walls are supported on beams double joists or blocking, and D) braced wall line spacing does not exceed twice the building width measured parallel to the braced wall line.
[3] Buildings are limited to two-story in SDC D_2, therefore anchor bolts at 4'-0" on center spacing do not apply.

Table 3-2 Summary of 2003 *IRC* Continuous Foundation and Anchor Bolt Requirements for Bearing Walls in One- and Two-family Houses

Seismic Design Category	Continuous Foundation Required At Bearing Walls	Anchor Bolts Required	Anchor Bolt Diameter Spacing
A, B and C	YES along Exterior Walls	YES	½" @ 6'-0"
A, B and C	NO along Interior Bearing Walls	NO	N/A
D_1	YES along Exterior Walls	YES[1]	½" @ 6'-0" ½" @ 4'-0" >2 story
D_1	YES along Interior Bearing Walls supported on a slab-on-grade	YES[1]	½" @ 6'-0" ½" @ 4'-0" >2 story
D_1	NO along Interior Bearing Walls supported on a raised floor	NO	N/A
D_2	YES along Exterior Walls	YES[1]	½" @ 6'-0" [2]
D_2	YES along Interior Bearing Walls supported on a slab-on-grade	YES[1]	N/A
D_2	NO along Interior Bearing Walls supported on a raised floor	NO	N/A

[1] Requires a square plate washer 3" x 3" x ¼" on each bolt.
[2] Buildings are limited to two-story in SDC D_2, therefore anchor bolts at 4'-0" on center spacing do not apply.

Chapter 4
FLOOR CONSTRUCTION

Woodframe floor systems and concrete slab-on-grade floors are discussed in this chapter. Although cold-formed steel framing for floor systems also is permitted by the *IRC*, it is not covered here; rather, the reader is referred to the AISI *Standard for Cold-Formed Steel Framing – Prescriptive Method for One- and Two-Family Dwellings* (AISI, 2001) for guidance. Also permitted but not discussed here are pressure-treated wood floor systems on ground; information on the use of these systems is provided in *IRC* Chapter 5.

4.1 GENERAL FLOOR CONSTRUCTION REQUIREMENTS

Woodframe floor systems form a horizontal diaphragm at each level where they occur and transfer earthquake lateral loads to braced walls below that floor level or directly to the foundation when the lowest floor is supported on a foundation. When a floor supports walls above and is supported on walls below as shown in Figure 4-1, the lateral loads in the floor system are based on the mass of the floor itself and a portion of the mass of all the walls in the stories immediately above and below the floor. (See Chapter 2 of this guide for a discussion of the complete load path.)

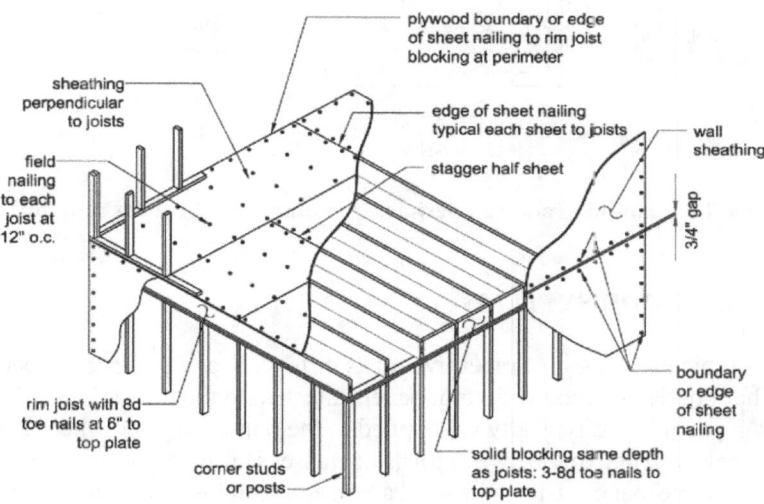

Figure 4-1 An unblocked floor diaphragm.

Concrete slab-on-grade floors typically are constructed with a concrete perimeter foundation and together these elements form the base of the building. Lateral forces from exterior and interior braced wall lines are transferred to a slab-on-grade via connections between the bottom plate of a braced wall and the slab. In turn, the concrete slab and foundation transfer those forces directly to the ground as shown in Figure 4-2. (For more information on anchorage of braced walls to slab-on-grade construction, see Chapter 3 of this guide.)

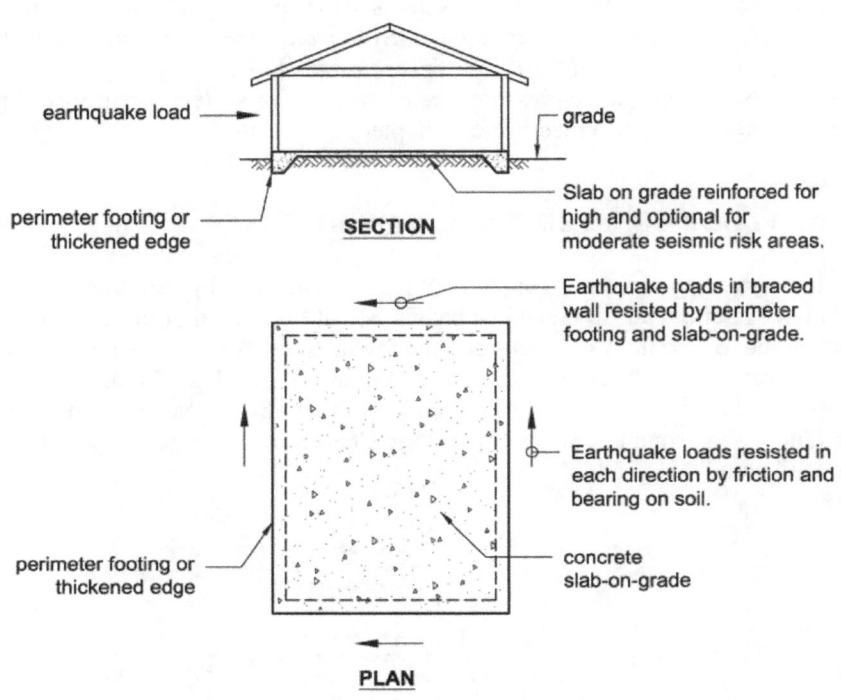

Figure 4-2 Slab-on-grade and perimeter footing transfer loads into soil.

4.2 WOODFRAME FLOOR SYSTEMS

Woodframe floors typically consist of repetitive joists or trusses, at a prescribed spacing, sheathed with either boards or wood structural panels attached to the top surface. Finish materials such as gypsum board typically are applied to the bottom surface where it serves as the ceiling for a room below. Blocking between joists or trusses is used at the ends of the floor joists or trusses (or a continuous band joist can be used at the ends) and where walls occur above or below. Floor systems also include beams, girders, or headers where needed to support joists. Joists can be sawn lumber, end jointed lumber, or a variety of prefabricated (engineered) members. Examples of engineered lumber include wood I-joists, trusses, and solid rectangular structural composite members such as parallel strand lumber (PSL), laminated veneer lumber (LVL), and laminated strand lumber (LSL). Beams, girders, or headers and blocking also can be either sawn lumber or engineered lumber.

The primary design consideration in choosing the minimum size and the maximum span and spacing of floor joists, trusses, beams, girders, and headers is adequate support for dead and live vertical loads as prescribed by the code depending on the uses that a floor must support. Vertical deflection of a floor is another design consideration that can limit the maximum span of floor members. Tables in *IRC* Chapter 5 and similar tables in other documents such as those published by the American Forest and Paper Association (AF&PA) or engineered lumber manufacturers are available for use in selecting the proper combination of minimum size and maximum span and spacing of floor framing members.

4.3 CANTILEVERED FLOORS

When floor joists cantilever beyond a support, joist size and spacing are limited by prescriptive tables in *IRC* Chapter 5. *IRC* Table R502.3.3(1) addresses support of a roof and one story of wall for roof spans up to 40 feet and snow loads up to 70 psf. *IRC* Table R502.3.3(2) addresses cantilever joists supporting an exterior balcony. When a floor is supporting more than a roof and one story of wall, the maximum prescriptive cantilever distance is limited by the *IRC* to the depth of the joist. If longer cantilevers are desired, a registered design professional must design that portion of the floor system.

In Seismic Design Categories D_1 and D_2, when cantilevered floor joists support braced wall panels in the story above, the cantilevered floor is limited by several additional prescriptive requirements in *IRC* Chapter 3. This is because the braced wall above and braced wall below are offset out-of-plane. When a floor cantilever supporting a braced wall does not meet the *IRC* limits, that portion of the house is defined as having an irregularity that prevents the use of prescriptive wall bracing where the irregularity occurs. In such a case, engineering must be applied to resolve the out-of-plane offset of the braced walls located in the stories above and below that floor. The maximum permitted cantilever of a second floor supporting a braced wall and roof is illustrated in Figure 4-3. (Also see the discussion of load path in Chapter 2 of this guide.)

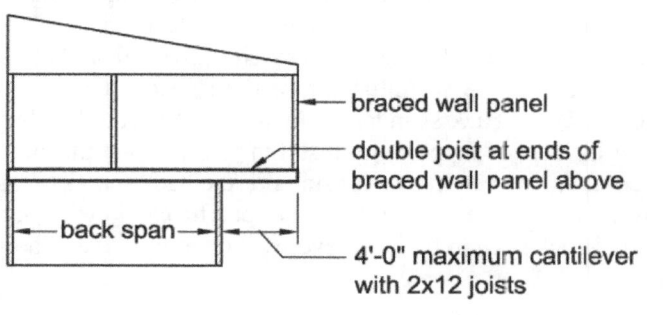

Figure 4-3 Cantilevered floor restrictions.

The specific limits and requirements in *IRC* Chapter 3 for cantilevered floors in SDCs D_1 and D_2 that support braced walls are not particularly difficult to meet and appear to omit addressing the uplift restraint that may be necessary at the back span support of cantilever joists. In SDCs D_1 and D_2, cantilever floor joists supporting a braced wall panel may not extend more than four times the nominal depth of the joist when the following set of rules is met:

- Joists must be 2x10 nominal or larger at 16-inch maximum spacing.

- The back span of the cantilever joist must be at least twice the cantilever distance.

- Joists must be doubled at the ends of the braced wall panel above.

- A continuous rim joist is connected to the end of each cantilevered joist. If that rim joist is spliced along its length, the splice must be made with either: (a) a 16-gage strap having 6 – 16d common nails on each side of the splice or (b) by using wood blocking having the same size as the rim joist, installed between the cantilevered joists, and nailed to the rim with 8 –16d common nails on each side of the splice.

- The cantilever end of the joist is limited to supporting uniform loads from a roof and wall above and, if supporting a header above, the header span is limited to 8 feet.

These rules are illustrated in the framing plan shown in Figure 4-4. What is not mentioned in this list of rules is the need for connections to resist uplift at the back-span (interior) end of a cantilever joist as noted in Figure 4-4. In *IRC* Table R502.3.3(1) for cantilever joists supporting a roof and wall, the uplift is determined using a back-span distance that is three times the cantilever distance (3:1). Because the minimum back span specified in the *IRC* Chapter 3 (see Item 2 above) is only twice the cantilever distance (2:1), the uplift values in *IRC* Table R502.3.3(1) would need to be increased by a factor of 1.5 just to address the gravity loads.

When the downward earthquake overturning load from the ends of a braced wall panel supported by cantilever joists are considered in addition to gravity loads, the uplift load at the back-span end of the joist obviously will increase. Therefore, depending on the actual back-span-to-cantilever-length ratio, the back-span end of the double cantilever joists supporting the ends of a braced wall may need to provide uplift restraint as much as twice than that listed in *IRC* Table R502.3.3(1). However, because the magnitude of the uplift load at the back-span end of a cantilevered joist reduces as the back-span length increases, it is possible that a cantilever joist that is continuous over its interior support will result in zero uplift at the back-span end. When cantilever joists are continuous over an interior support, the back span increases and the uplift at the end of the joist is greatly reduced. Therefore, the specific cantilever floor joist layout and ratio of the length of the back span to the cantilever will determine if and how much uplift may need to be resisted.

> **Above-code Recommendation:** In SDCs C, D_1 and D_2, when a braced wall is supported at the ends of cantilever joists, the back-span uplift connection capacity should be determined using engineering principals for the specific back-span and cantilever distances involved.

Chapter 4, Floor Construction

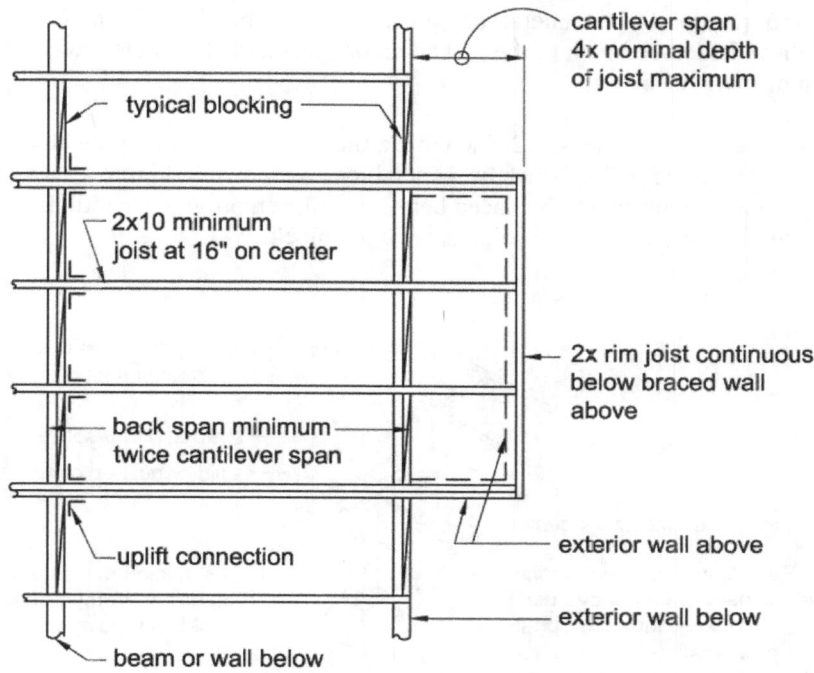

Figure 4-4 Cantilever joist at braced wall above.

4.4 REQUIREMENTS FOR BLOCKING

It is important in floor framing construction to prevent joists (or trusses) from rotating or displacing laterally from their intended vertical position. Rotation loads occur because, when floor sheathing is resisting lateral loads oriented perpendicular to the joist, those lateral loads are actually trying to move the top edge of the joist sideways.

Preventing rotation is often accomplished by installing full-depth solid blocking at the ends of joists. The ends of joists also can be restrained by attaching the joist to a continuous rim or band joist or a header or, in balloon framed walls, by attaching the joist to the side of a stud. In SDCs D_1 and D_2, additional solid blocking between joists (or trusses) is necessary at each intermediate support even when that location is not at the end of the joist. For example, intermediate support should be located at an interior girder or bearing wall where joists are continuous over that support. Blocking installed between joists supported by an interior floor girder is illustrated in Figure 4-5.

Blocking also is required below an interior braced wall line in all Seismic Design Categories when joists are perpendicular to the braced wall. Although the *IRC* is silent regarding minimum depth and width for these blocks, the intent of this added blocking is to provide a nailing surface for the 16d common nails used to connect the bottom plate of the braced wall to the floor. This nailing is an important part of the lateral load path; therefore, the blocking should be of a depth sufficient to allow full embedment of the 16d common nails and of sufficient width to prevent the nails from missing the block.

Assuming floor sheathing is at least 1/2- inch thick, the minimum depth of the blocking should be 1-1/2 inches. Therefore, a flat 2-inch by 4-inch block as shown in Figure 4-6 can provide sufficient depth and, when accurately placed below a wall, can provide a width that greatly reduces the potential for bottom plate nails missing the block.

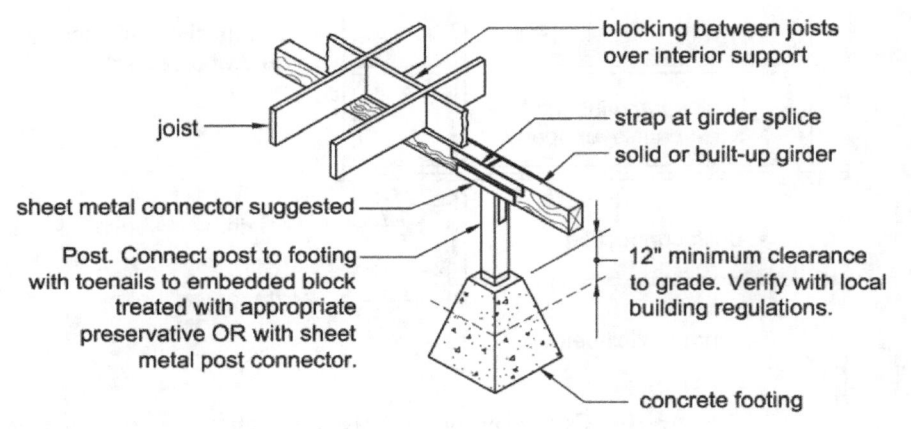

Figure 4-5 Interior bearing line.

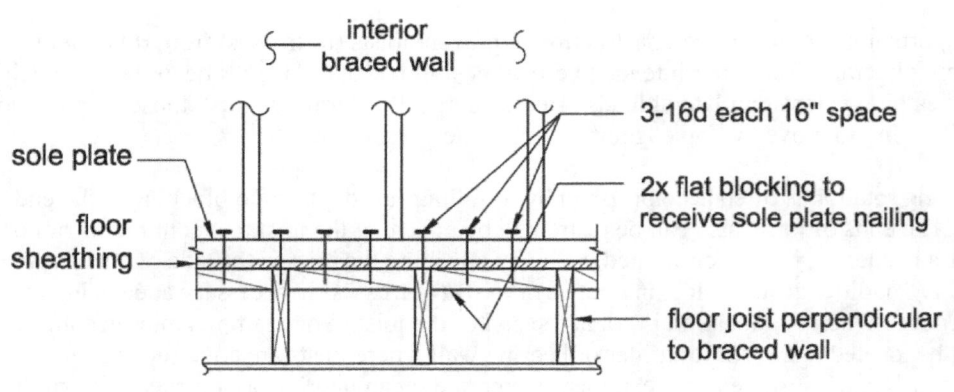

Figure 4-6 Blocking below interior braced wall where floor joists are perpendicular to wall.

When an interior braced wall also is a bearing wall and joists below the wall are parallel to the wall, a double joist or a beam typically is provided in the floor below the wall. Occasionally this pair of joists may be spaced apart to allow for piping or vents passing vertically from the wall above through the floor. When this occurs, the double joists cannot be located directly below the wall's bottom plate. To provide a nailing surface for the bottom-plate connection of the braced wall above, 2x flat blocking should be installed in line with the braced wall's bottom plate between and parallel to these spaced joists as shown in Figure 4-7.

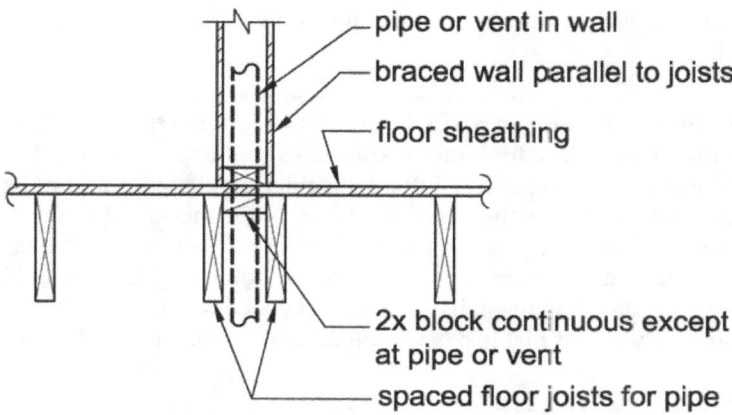

Figure 4-7 Blocking for floor joists spaced apart for piping in floor.

4.5 CONNECTION OF FLOOR JOISTS TO WALL TOP PLATE OR FOUNDATION SILL PLATE BELOW

Floor joists (or trusses) are required to be connected to the top plate of supporting walls or to a foundation sill plate as specified in *IRC* Table R602.3(1). Each of these connections provides a load path to transfer loads from the floor diaphragm into the braced walls or the foundation below. Nailed connections must meet the following minimum requirements:

- Rim or band joists parallel to a wall or foundation require a toe-nailed connection to the wall top plate or foundation sill plate using 8d box or common nails at 6 inch spacing.

- Floor joists perpendicular to a wall top plate or foundation sill plate require a toe-nailed connection using three 8d box or common nails.

- When blocking is installed between the floor joists, the blocking requires a toe-nailed connection to the wall top plate or foundation sill plate using a minimum of three 8d box or common nails in each block.

Where toe nailing is used, it must be done correctly so that it can transfer the intended loads and so that the nails do not split the wood when being installed. Toe-nailed connections prescribed by the *IRC* should be acceptable when connecting joists to wall top plates or foundation sills that are perpendicular to the joists because these connections are not highly loaded by lateral loads. The primary lateral load transfer in a floor system occurs through the rim or band joists and through blocking that is parallel to braced walls or foundation sill plates.

Information on proper toe-nail installation is presented in Figure 4-8; however, that idealized picture of nail inclination and location is difficult to achieve in actual construction. Consequently, many toe-nailed connections that must transfer lateral loads may not actually perform very well.

> **Above-code Recommendation: In SDCs C, D_1 and D_2, connections between joists or blocking and wall top plates or foundation sill plates that are parallel to the joist or blocking should use commercially available light-gage steel angles and nails of the correct diameter and length for the product.** Many of the toe-nailed connections specified in the *IRC* also can be made using light-gage steel angles through which face nails are driven into the two wood framing members being connected. Although light-gage angles may require more time to install than toe nails, the angle connections should reduce splitting of the wood and can provide a more reliable connection capacity for lateral loads compared to toe nails.

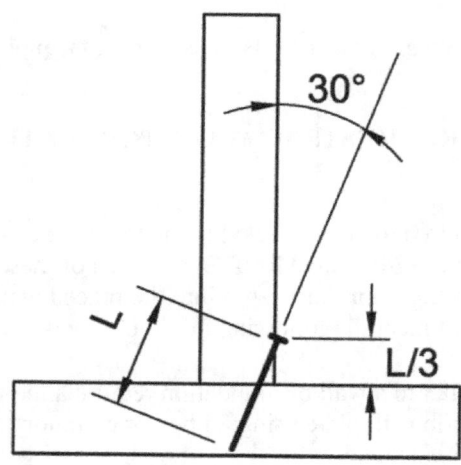

Figure 4-8. Toe nail configuration requirements.

4.6 FLOOR SHEATHING

Floor sheathing can be either wood boards installed perpendicular or diagonally to the joists or wood structural panels (as subfloor or combination subfloor-underlayment) installed with the long direction of the panel perpendicular to the joists. The minimum thickness for wood board sheathing depends on joist spacing and the orientation of the boards to the joist (e.g., perpendicular or diagonal). For wood structural panels, the minimum thickness is based on joist spacing and the grade of sheathing panels selected. *IRC* Chapter 5 contains tables to use in determining the required minimum thickness for sheathing materials based on a variety of joist spacings.

> **Above-code Recommendation:** Wood boards are rarely used in modern house construction unless the underside of the floor is intended to be visible to the interior space below to achieve a specific architectural effect. **As an above-code measure, wood boards installed perpendicular to joists should not be used in SDC C, D_1, or D_2 unless wood structural panels are installed over the boards because the wood board sheathing alone provides little resistance to lateral loads.** In contrast, diagonally placed wood boards provide greater lateral capacity and should be acceptable for small rectangular-shaped floor areas.

For modern construction, floor sheathing typically will be wood structural panels (OSB or plywood). These panels are fastened to the joists based on a schedule prescribed in tables in *IRC* Chapter 6.

4.7 LATERAL CAPACITY ISSUES FOR WOOD FRAMED FLOORS USING WOOD STRUCTURAL PANELS

The lateral capacity of a floor diaphragm sheathed with wood structural panels is based on five factors:

- Sheathing thickness,
- Sheathing fastener type and size,
- Fastener spacing along supported sheathing edges,
- Presence or absence of blocking along all edges of each piece of sheathing, and
- Layout of the sheathing joints with respect to direction of lateral loading.

Below is a discussion of how differences in lateral capacity can result depending on how each of these is applied to the construction of a floor.

Sheathing thickness usually is selected based on the spacing of joists and, for floors, will never be less than 7/16 inch but normally is at least 5/8 inch. Generally, thicker sheathing will provide a more comfortable floor for the occupants to walk on and will have a greater lateral capacity compared to thinner sheathing using the same fastener size and spacing.

The most common sheathing fasteners used are nails with a minimum size of 6d common (0.113 inch x 2 inches) for a floor sheathing thickness of up to 1/2 inch. The minimum fastener size increases with increasing sheathing thickness to a minimum of 10d common nails (0.148 inch x 3 inches) for sheathing that is 1-1/8 inches thick. Larger diameter nails will provide greater lateral capacity than smaller nails in the same thickness of sheathing because the lateral capacity of a nail is directly proportional to its diameter. Therefore, using box nails or gun nails that have a smaller diameter than common nails will reduce the lateral capacity of a floor diaphragm.

Staples also can be used to fasten sheathing to framing members. Although not commonly used, *IRC* Table R602.3 (2) has information for specifying alternative sheathing fasteners including staples. Generally, staples of either 15 or 16 gage can be used in place of most nails at the same spacing as those nails. However, when using staples, it is important to understand that they must be installed with the crown parallel to the framing member below the sheathing edge being fastened.

Fastener spacing for floor sheathing is typically 6 inches along continuously supported panel edges and 12 inches along supporting members not located at panel edges. Greater lateral capacity can be obtained when fastener spacing along supported edges is reduced from 6 inches to 4 or 3 inches.

The *IRC* only requires floor diaphragms to be fastened along continuously supported panel edges. This includes where panel edges are located parallel to and over a joist and at the floor framing members forming the perimeter of the floor. The unsupported panel edges that are spanning perpendicular to the joists only need to be fastened at each joist. In engineering terms, this is called an unblocked diaphragm. See Figure 4-1 for the sheathing layout and nailing pattern for a portion of an unblocked diaphragm.

In contrast, a fully blocked floor diaphragm means that all edges of each sheathing panel that are not located on a joist are supported on and fastened to blocking. A blocked diaphragm will have significantly greater lateral capacity than an unblocked diaphragm having the same thickness of sheathing and attached with identical fasteners because the extra fasteners along the blocked edges provide additional capacity. Figure 4-9 shows the layout of sheathing and nailing of a portion of a blocked floor diaphragm. Fastening the sheathing to joists or blocking along all panel edges allows the shear loads being carried in the sheathing to be transferred from one panel to the next much more effectively. This, in turn, ties the floor together better and allows the braced walls below that floor to resist an earthquake more as a system than as individual walls.

> **Above-code Recommendation:** The *IRC* requires that wood structural panel sheathing be installed with the long dimension of the panel perpendicular to joists, but it does not specify staggering of panel joints along the short direction of the panels. **Although not specifically required by the *IRC*, sheathing should be installed as shown in Figures 4-1 and 4-9 to achieve the greatest capacity**. This staggered sheathing layout pattern causes the individual sheathing panels to interlock and makes the whole floor act as a unit.

Chapter 4, Floor Construction

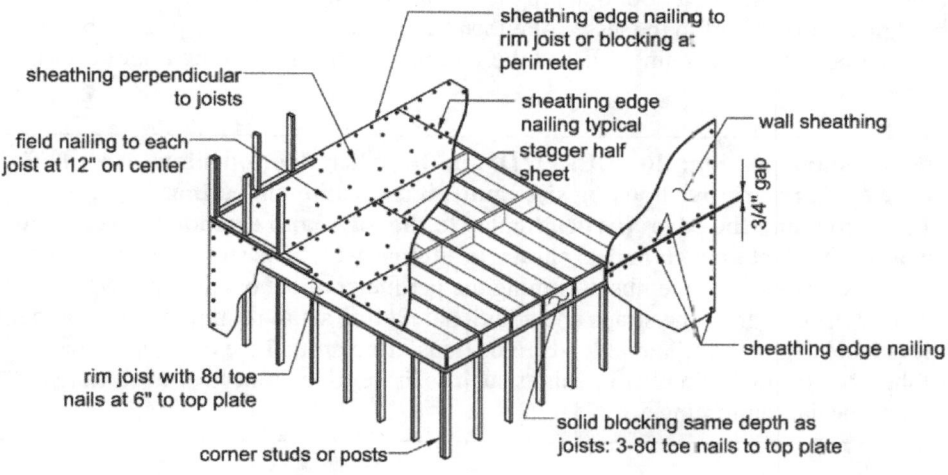

Figure 4-9 Blocked diaphragm configuration.

Lateral loads in a floor diaphragm also are affected by the distance between braced wall lines or between foundations located below the floor. The loads increase with increasing distance between lines of parallel braced walls or foundations. Therefore, a long and narrow floor diaphragm as shown in Figure 4-10 will have to transfer a greater load per foot along its short sides than along its long sides. In order to limit the maximum load along a short side, *IRC* Chapter 6 places limits on the maximum spacing between braced wall lines or foundations.

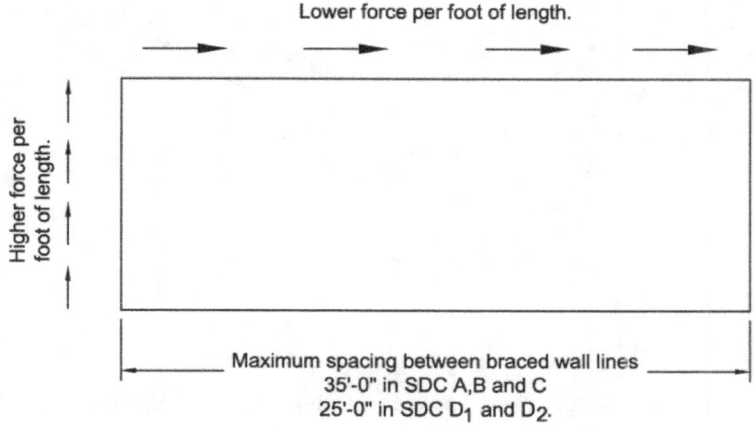

Figure 4-10 Diaphragm loads on long and short sides.

79

The size and location of floor openings such as for stairs or a two-story entry foyer can create concentrations of lateral loads in a floor diaphragm. To address this, *IRC* Chapter 3 limits openings through a floor or roof to the lesser of either 12 feet maximum or 50 percent of the least dimension of the floor. When openings exceed these limits, engineering of the floor or roof diaphragm is required.

> **Above-code Recommendation:** In SDCs C, D_1 and D_2, when floor openings exceed 50 percent of the *IRC* prescriptive opening size limits, it is recommended that 16-gage straps be installed along the edges perpendicular to the joists and extended beyond the opening by at elast 2 feet at each end as shown in Figure 4-11. The straps can be nailed with 10d nails into the framing members forming the perimeter of the opening and into blocking beyond the corners. The straps and additional nailing act to reinforce the diaphragm and provide a dedicated path for lateral loads to be transferred around the opening to the portions of the floor beyond. Smaller openings such as those for chimneys or duct shafts do not require any special reinforcing.

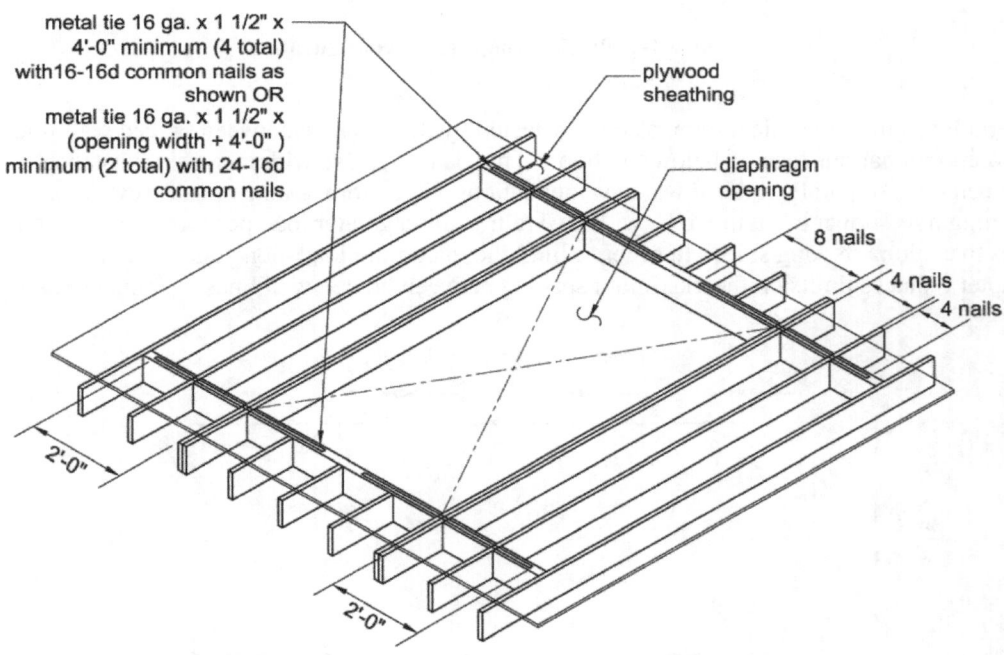

Figure 4-11 Reinforcing straps at large diaphragm openings.

4.8 CONCRETE SLAB-ON-GRADE FLOORS

A concrete slab-on-grade can be used as the base of a first-floor level or of a basement level. The minimum thickness for a concrete slab-on-grade is 3-1/2 inches except where expansive soil is present. Where expansive soils are encountered, a design for the slab-on-grade must conform to *IBC* Chapter 18 (see Chapter 3 of this guide for a discussion of the effects of expansive soil).

Concrete Strength Requirements – The minimum concrete compressive strength is 2,500 pounds per square inch (psi). Floor slabs having an exterior surface exposed to the weather in areas of moderate to severe concrete weathering must have higher compressive strength as specified in *IRC* Table R402.2. A map in *IRC* Chapter 3 identifies locations where moderate and severe weathering of concrete is expected to occur.

Reinforcing of Concrete Slab-on-Grade Floors – In the absence of expansive soils, the *IRC* does not require reinforcing of concrete slabs. Reinforcing typically is used to provide tension capacity in concrete and thereby reduce cracking caused by a variety of loads including temperature variations. Concrete alone has very good compression capacity but has a very low capacity for tension. Therefore, adding reinforcing bars to a slab-on-grade will provide much greater resistance to tension loads originating from earthquake loads and other soil conditions that could induce tension stress in the slab.

Reinforcing is required only where a slab-on-grade is thickened along its perimeter edge or below an interior bearing wall in SDCs D_1 and D_2. When exterior braced walls are spaced more than 50 feet apart, an interior braced wall also needs a foundation as part of the slab-on-grade in SDCs D_1 and D_2. (See Chapter 3 of this guide for information on where and how much reinforcing is needed in foundations provided with a slab-on-grade.)

> **Above-code Recommendation:** When a slab-on-grade is placed separately from the perimeter footing below, the *IRC* does not specify any vertical reinforcing across this joint. When this condition occurs, the possibility of earthquake loads causing sliding along that joint is very real. **Therefore, as an above-code measure in SDCs C, D_1 and D_2, when the footing and the slab concrete are separately placed, it is recommended that vertical reinforcing dowels be placed across the joint between the slab and footing. These vertical dowels should be No. 4 steel reinforcing bars at 48-inch maximum spacing.** (More information on this subject is presented in Chapter 3 of this guide.)

Chapter 5
WALLS

5.1 WOOD LIGHT-FRAME CONSTRUCTION

5.1.1 General Components

In residential construction, the walls provide the primary lateral resistance to wind and earthquake loads. Even in frame type houses (e.g., post-beam construction), the exterior walls provide most of the lateral stability to the house. Although this guide focuses on wood light-frame construction, alternatives such as cold-formed steel, masonry, and concrete construction are used in many regions of the country. The reader is referred to the sections on masonry and concrete construction later in this chapter for some guidance on the use of these materials. For cold-formed steel construction, the reader is referred to the American Iron and Steel Institute's (AISI) industry standard for prescriptive cold-formed steel construction, *Standard for Cold-formed Steel Framing Prescriptive Method for One- and Two-Family Dwellings* (2001).

Light-frame walls provide resistance to sliding, overturning, and racking loads induced in the house by an earthquake as illustrated in Figure 5-1. The walls are the principle element for transmitting the loads from the upper stories and roof to the foundation. The concept of how these loads are transferred between the major components of the house is illustrated in Figure 5-2, and the action of the individual wall segments resisting the lateral loads is illustrated in Figure 5-3. Wood light-frame walls typically consist of the lumber framing covered by sheathing material that is attached to the wood framing with nails, staples, or screws. Figure 5-4 illustrates the components of a wall that is sheathed with wood structural panels (OSB or plywood) on the outside and gypsum wallboard on the inside. The figure also shows the addition of hold-down connectors to the framing, which are required by the *IRC* for some specific bracing configurations. When used, hold-down connectors increase the strength and stiffness of the wall segment.

Four different bracing wall configurations and eight methods (materials) are recognized by the *IRC*. The bracing wall configurations include:

- *IRC* Section R602.10.3 braced wall panels (Figure 5-5a),
- *IRC* Section R602.10.5 continuous (wood) structural panel sheathing (Figure 5-5b),
- *IRC* Section 602.10.6 alternate braced wall panels (similar to Figure 5-5c), and
- Wood structural panel sheathed walls with hold-down connections as required by the exceptions in *IRC* Section R703.7 when stone or masonry veneer is used (Figure 5-5c).

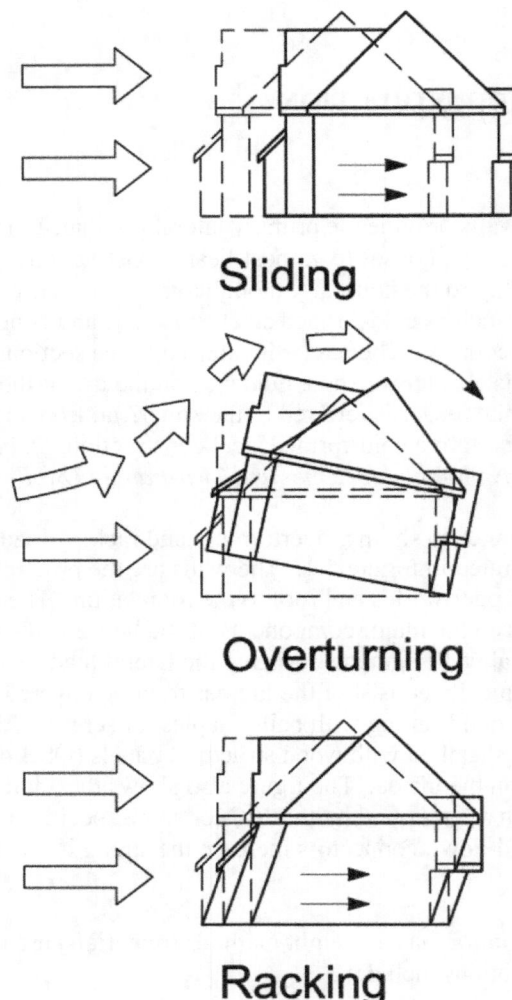

Figure 5-1 Sliding, overturning, and racking action resisted by walls and foundation.

Chapter 5, Walls

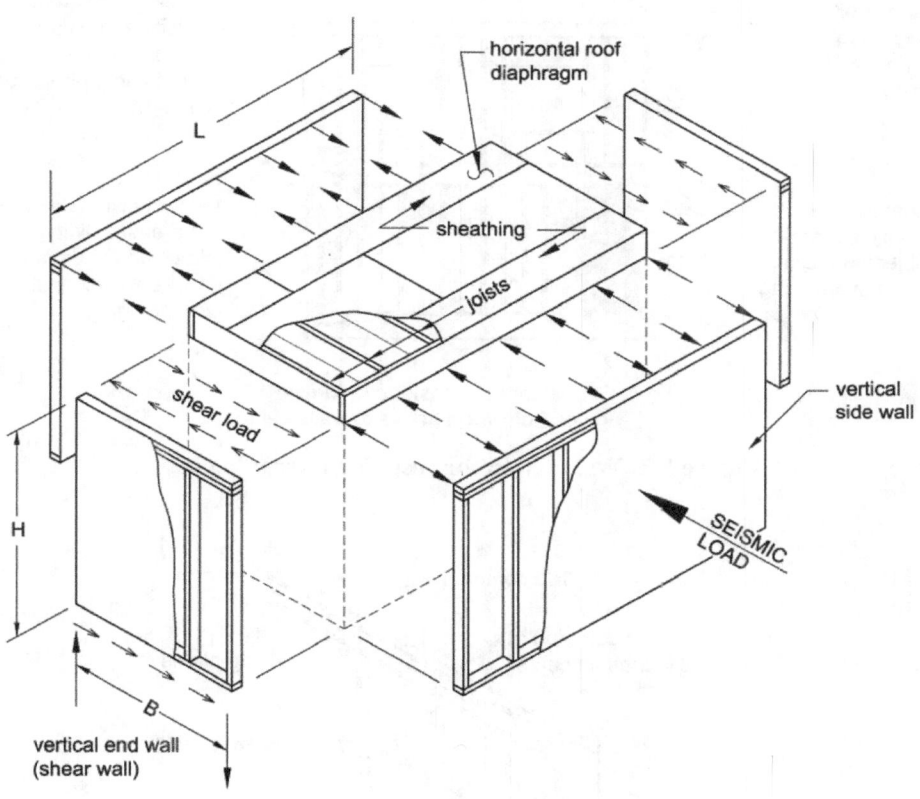

Figure 5-2 Exploded view of house illustrating load paths.

FEMA 232, Homebuilders' Guide

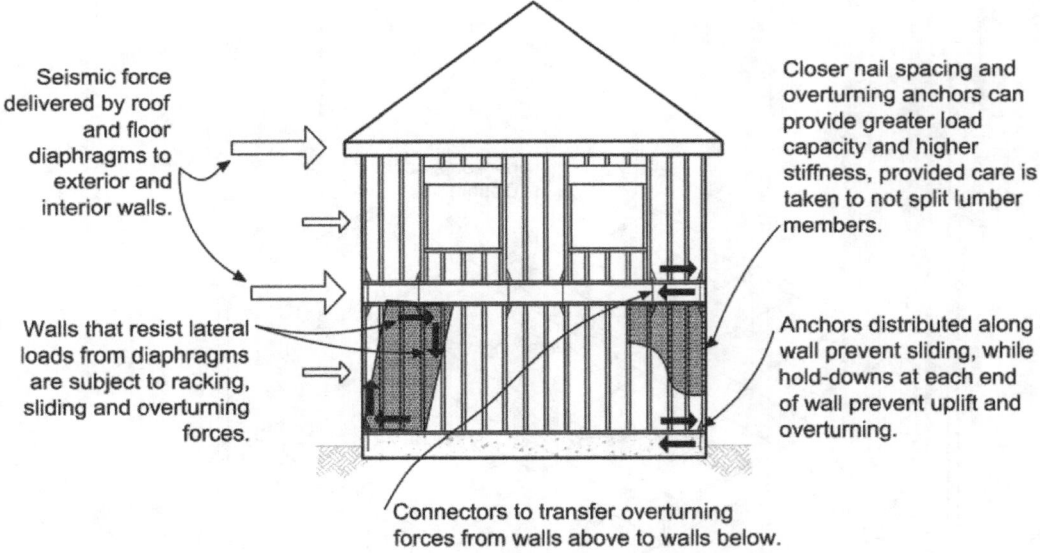

Figure 5-3 Wall action for resisting lateral loads

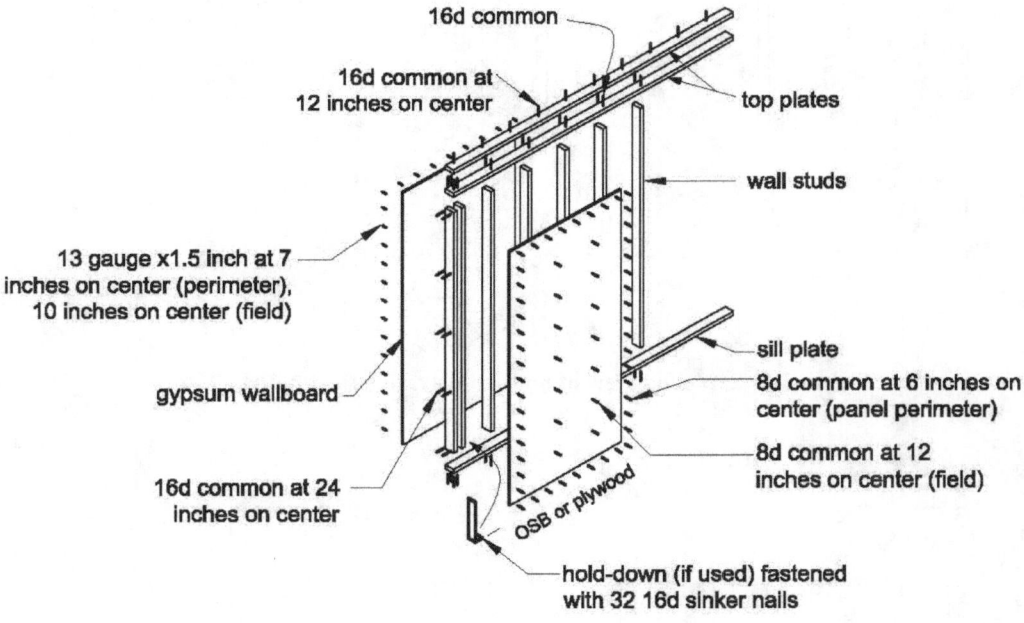

Figure 5-4 Exploded view of typical residential wall segment.

Differences in these bracing wall configurations include sheathing materials, minimum bracing length, extent of sheathing, and anchorage at the wall base. Differences in overturning anchorage for walls are shown in Figure 5-5.

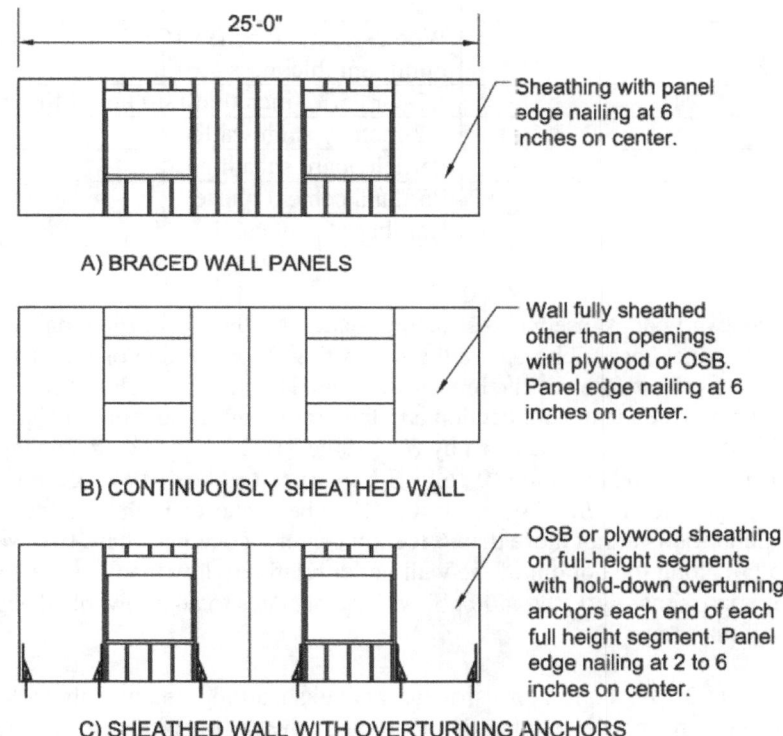

Figure 5-5 **Detailing differences for three options when using wood structural panel sheathed walls.**

The braced wall panel (*IRC* Section R602.10.3) is the most commonly used approach. Eight different methods (materials) are recognized by the *IRC* as acceptable bracing. These are called "braced wall panel construction methods" in the *IRC* and are listed in Table 5-1. Method 1, let-in bracing, is not allowed to be used in regions of high earthquake hazard because it often will fail as the walls are racked during an earthquake; therefore, this method is not discussed further in this guide.

Of the acceptable braced wall panel materials, wood structural panels and diagonal lumber sheathing are known to perform better than others (i.e., withstand higher deformations while supporting higher loads). Wall panel bracing is required by *IRC* Section R602.10.4 to be provided in 4-foot minimum lengths sheathed on one face for other than Method 5 and either 4-foot lengths sheathed on both faces or 8-foot lengths sheathed on one face for Method 5. Other than with Method 5, this bracing often is provided in 4-foot-long isolated segments along the wall length.

Table 5-1 Braced Wall Panel Construction Methods (Materials) Recognized by the *IRC*

Construction Method Designation	Sheathing Material
1	Nominal 1x4 inch continuous let-in bracing
2	5/8-inch minimum thickness boards applied diagonally to studs
3	Wood structural panels (OSB or plywood) 5/16-inch minimum thickness
4	1/2- or 25/32-inch thick structural fiberboard
5	1/2-inch gypsum wallboard
6	Particleboard sheathing
7	Portland cement plaster
8	Hardboard panel siding

Where braced wall panels use wood structural panel (Method 3) or diagonal lumber (Method 2) sheathing, the panel base anchorage to the supporting floor framing or foundation limits the bracing strength and stiffness. The braced wall panel anchorage includes two critical weak links for uplift: the panel end stud connection and the bottom plate (sole plate) connection. Braced wall panel bottom plates are specified by *IRC* Table R602.3(1) to be attached to the floor platform using three 16d common (0.162 x 3.5 inch) or 16d box (0.148 x 3.5 inch) nails every 16 inches or are specified by *IRC* Section R403.1.6 to be anchored to the foundation with 1/2-inch-diameter anchor bolts at not more than 6 feet on center. Together, these two weak links cause the wall to fail along the bottom of the wall under relatively low loads (these walls have a capacity of approximately 150 to 400 plf, which translates to an allowable design value of about 50 to 140 plf maximum).

IRC Section R602.10.5, continuous structural panel sheathing, requires that all exterior wall surfaces on a given story level other than door and window openings be sheathed with wood structural panel sheathing (Figure 5-5b). This bracing wall configuration has greater strength and stiffness than braced wall panels with minimum wall base anchorage. The increased strength and stiffness are due in part to the continuity provided by additional sheathing above and below windows and doors (which makes the wall-base connection capacity less critical) and increased overturning capacity due to required corner framing details or hold-downs at wall ends. In recognition of the improved strength and stiffness provided with continuous structural panel sheathing, *IRC* Section R602.10.5 permits the minimum length of bracing to be reduced and *IRC* Table R602.10.5 permits use of individual wall bracing panels that are more slender than would otherwise be permitted.

Wood structural panel wall bracing with uplift anchorage provided at each end, per *IRC* Sections R602.10.6 and R703.7, is the strongest and stiffest option for resisting lateral loads. This wall configuration is illustrated in Figure 5-5c. Although sheathing, fastening, and hold-down loads for these walls are prescribed, the walls are essentially equivalent to engineered shear walls. The alternate braced wall panel provisions of *IRC* Section R602.10.6 were developed to allow use of braced wall panels narrower than the 4-foot minimum required by *IRC* Section R602.10.4 (e.g., at the side of garage doors); however, the *IRC* permits an alternate braced panel to be substituted for each 4 feet of bracing throughout the house. The provisions of *IRC* Section R703.7 use increased wall strength and stiffness to compensate for the increased earthquake loads that occur due to the weight of stone or masonry veneer.

As discussed above, where wood structural panel bracing is used, the strength and stiffness of the wall bracing is very dependent on the extent of sheathing and anchorage at the wall base. The configuration shown in Figure 5-5a, using *IRC* minimum braced wall panel anchorage, is the weakest and least stiff of the wood structural panel bracing configurations. The configuration shown in Figure 5-5b adds strength and stiffness by providing continuous structural panel sheathing and added detailing or tie-downs at wall ends. The strongest and stiffest configuration is illustrated in Figure 5-5c where overturning anchors are provided at each end of each wall segment.

> **Above-code Recommendation: Use of the configurations shown in Figures 5-5b or 5-5c can significantly increase the strength and stiffness of braced wall panels sheathed with wood structural panel or diagonal lumber sheathing and may be used to provide improved performance whether or not specifically required by the *IRC*.** Use with sheathing materials other than Methods 2 or 3 would provide less benefit and is not recommended.

When the wood structural panel wall bracing option is used in an engineered design, the walls are designed according to empirical tables that provide allowable design loads in pounds per foot of wall depending on the thickness and grade of the sheathing and the size and spacing of the sheathing nails. These walls rely on hold-down anchor connections to resist the overturning loads and substantial connections (nails, lag screws, bolts, etc.) along the top and bottom plates to transmit the lateral loads between the wall framing and the floor platform or foundation. This type of wall can resist allowable design loads up to 870 plf when sheathed on one face with wood structural panels.

5.1.2 Design Requirements

The 2003 *IRC* requires that wall bracing be provided both at exterior walls and at interior braced wall lines. *IRC* Section R602.10.1.1 specifies that interior braced wall lines must be added such that the distance between braced wall lines does not exceed 35 feet; however, allowance is made for spacing up to 50 feet. *IRC* Section R602.10.11 reduces the maximum spacing to 25 feet in SDCs D_1 and D_2. Within each braced wall line, *IRC* Table R602.10.1 specifies the minimum bracing as a percentage of the length of the wall line based on the wind and earthquake exposure and the story level under consideration (the lower in the house, the more sheathing is required to resist the higher loads). Some interpretations of the *IRC* would allow a house located in SDC C to be braced according to the amounts required for SDCs A and B due to the exemption for detached houses in SDC C from all earthquake requirements.

> **Above-code Recommendation: Houses in SDC C should be braced according to the requirements of SDC D_1.** Use of the required percentage of wall bracing in each 25-foot length of wall will provide a distributed resistance system of walls rather than a concentrated wall system. Experience has shown that a distributed wall system performs better in an earthquake than concentrated walls in only a few locations.

Table 5-2 excerpts a portion of this bracing information. As previously noted, the amount of bracing may be modified in accordance with *IRC* Section R602.10.5 when continuous wood structural panel sheathing is provided. In addition, an alternate braced wall panel (per *IRC* Section R602.10.6) can be substituted for each 4 feet of braced wall panel. The sheathing must be located at the ends of each braced wall line at least 25 feet on center. The idea is to provide a distributed resisting system rather than to concentrate the resistance in a few highly loaded wall segments. Overall, this provides a more robust structure that can resist the loads induced into the house close to the source of the load and not require the floor and roof diaphragms to transmit the loads long distances. Let-in bracing is allowed only for the top story of houses in SDC C due to its tendency to fail at relatively low lateral load levels.

Table 5-2 *IRC* **Sheathing Requirements for Seismic Design Categories C, D_1, and D_2**

Seismic Design Category	Floor Level	Sheathing Requirements
C	Top story	16% of wall line for wood structural panel and 25% for all other sheathing types
C	First story of 2-story or second story of 3-story	30% of wall line for wood structural panel and 45% for all other sheathing types
C	First story of 3-story	45% of wall line for wood structural panel and 60% for all other sheathing types
D_1	Top story	20% of wall line for wood structural panel and 30% for all other sheathing types
D_1	First story of 2-story or second story of 3-story	45% of wall line for wood structural panel and 60% for all other sheathing types
D_1	First story of 3-story	60% of wall line for wood structural panel and 85% for all other sheathing types
D_2	Top story	25% of wall line for wood structural panel and 40% for all other sheathing types
D_2	First story of 2-story or second story of 3-story	55% of wall line for wood structural panel and 75% for all other sheathing types
D_2	First story of 3-story	Not allowed prescriptively must use *IBC* design methods
D_2	Cripple walls	75% of wall line using wood structural panel sheathing only

Although the basic *IRC* bracing concept is reasonably straightforward, a number of other adjustments may modify the required length of bracing. Close attention is required to ensure that the *IRC* requirements are met. *IRC* Section R301.2.2.2.1 limits the dead load of assemblies for houses in SDCs D_1 and D_2. The maximum permitted for roof plus ceiling dead load is 15 psf (typical asphalt shingle roofs with gypsum ceilings); however, *IRC* Table R301.2.2.2.1 permits this assembly weight to be increased to 25 psf (for heavier roofing materials) provided that the length of wall bracing is increasd as specified. Footnote d to *IRC* Table R602.10.1 notes that the earthquake bracing requirements are based on 15 psf (exterior) wall dead load and permits the required earthquake bracing length to be multiplied by 0.85 for walls with dead loads of 8 psf or less, provided that the length is not less than required by *IRC* Section R602.10.4 (4 feet for all but let-in bracing and gypsum board, which require a length of 8 feet) nor less than required for wind loading.

In addition to the adjustments to required bracing length, the second paragraph of *IRC* Section R602.10.11 (errata, second printing) permits the wood structural panel wall sheathing to begin up to 8 feet from the corner in SDCs D_1 and D_2 provided that:

- A 2-foot braced panel is applied in each direction at the house corner or
- Tie-downs are provided at the end of the braced wall panel closest to the corner.

This is more stringent than for low SDCs where braced wall panels are permitted to be located up to 12 feet 6 inches from wall corners without providing tiedowns or sheathed corners and farther from corners if collectors are provided to carry loads to the braced wall panels. However, a house that is tied together at the corners will resist the loads expected from an earthquake much better than if the corners are not connected well. When using bracing other than wood structural panels in SDCs D_1 and D_2, braced wall panels must be located at each end of braced wall lines.

> **Above-code Recommendation: Braced wall panels should extend to every corner.**

IRC Section R703.7 presents additional bracing modifications where exterior veneer is used. These requirements are discussed in Section 5.2 of this guide. For the guide's model house, Figures 5-6a through 5-6c provide plan views identifying the bracing required when the house is in SDC C, has a light finish system such as vinyl siding, and has a crawl space. The bracing requirements for the same house located in SDC D_2 are shown in Figures 5-7a through 5-7c. Notice the significant increase in designated wall bracing due to the higher level loads expected in SDC D_2.

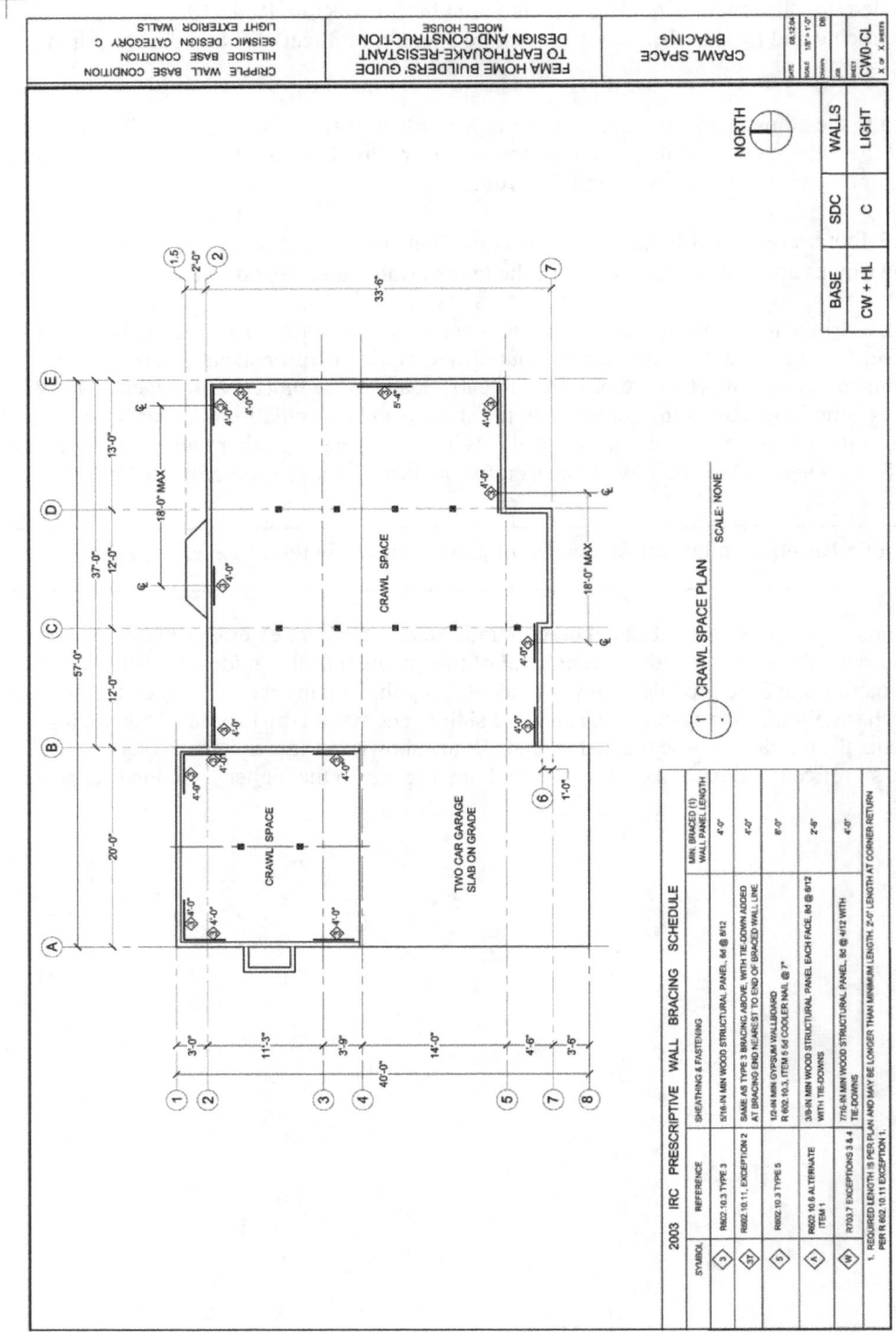

Figure 5-6a Plan view of crawl space for model house with light-weight finish material located in SDC C.

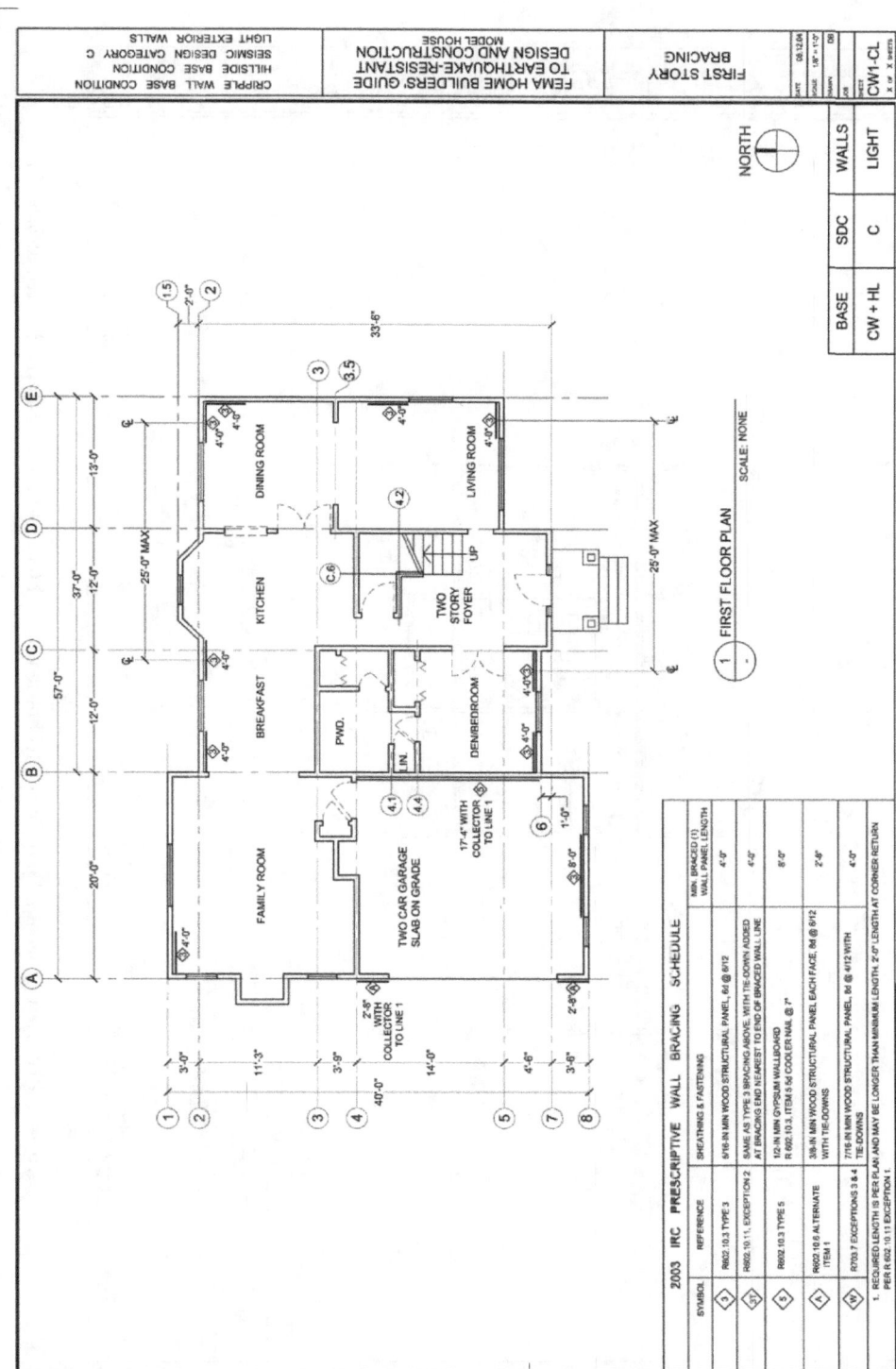

Figure 5-6b Plan view of first floor for model house with light-weight finish material located in SDC C.

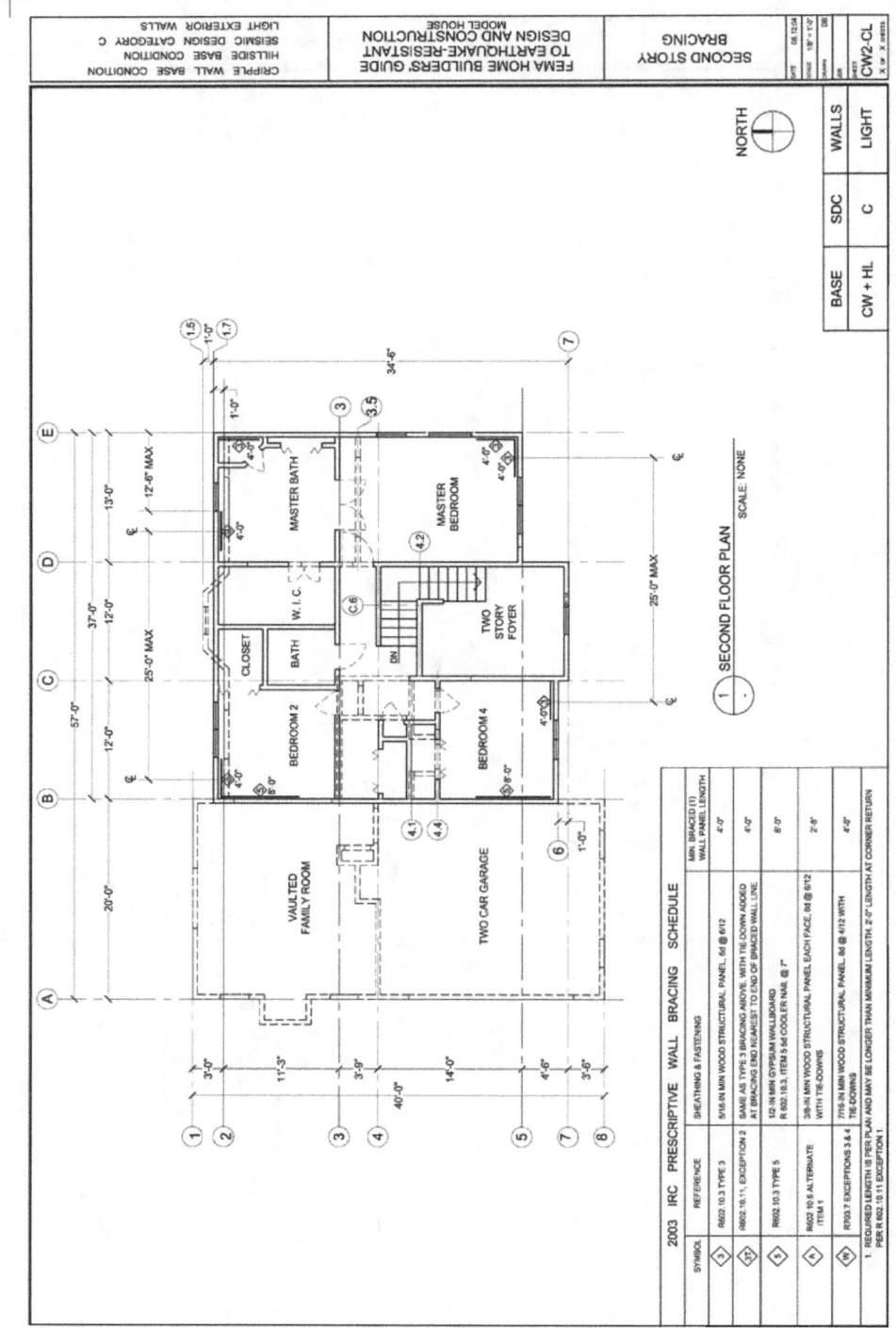

Figure 5-6c Plan view of second floor for model house with light-weight finish material located in SDC C.

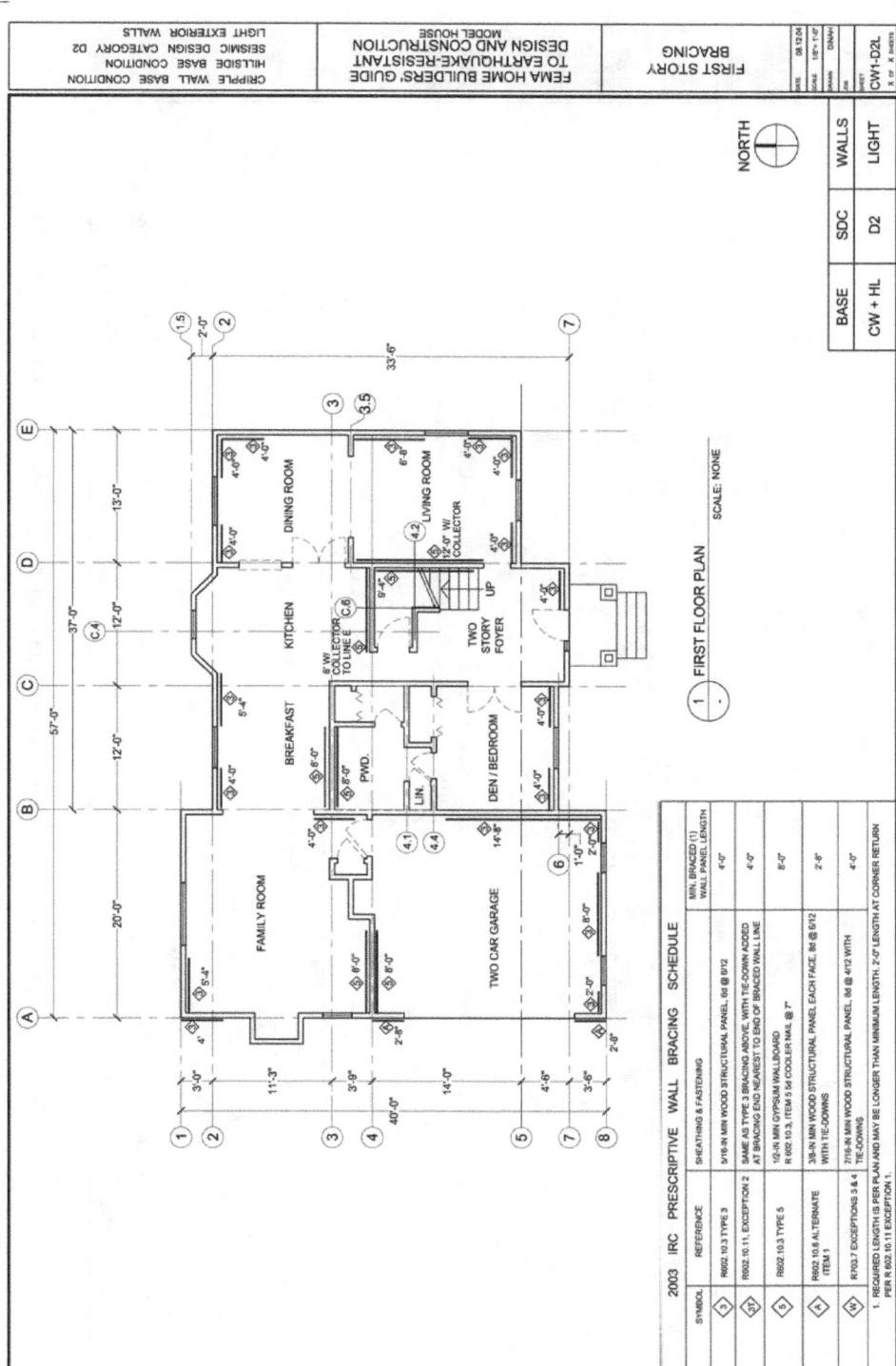

Figure 5-7a Plan view of crawl space for model house with light-weight finish material located in SDC D_2.

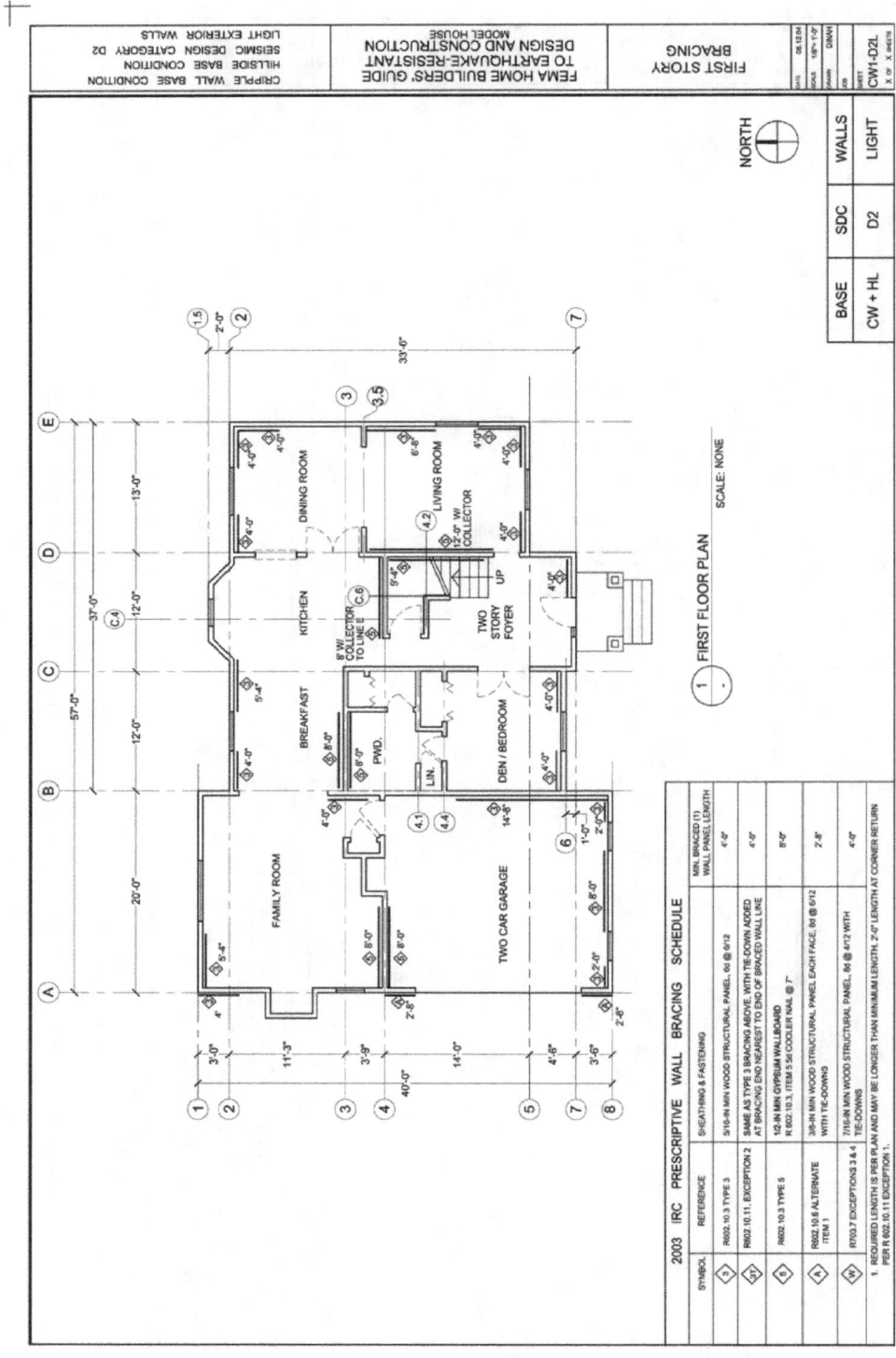

Figure 5-7b Plan view of first floor for model house with light-weight finish material located in SDC D_2.

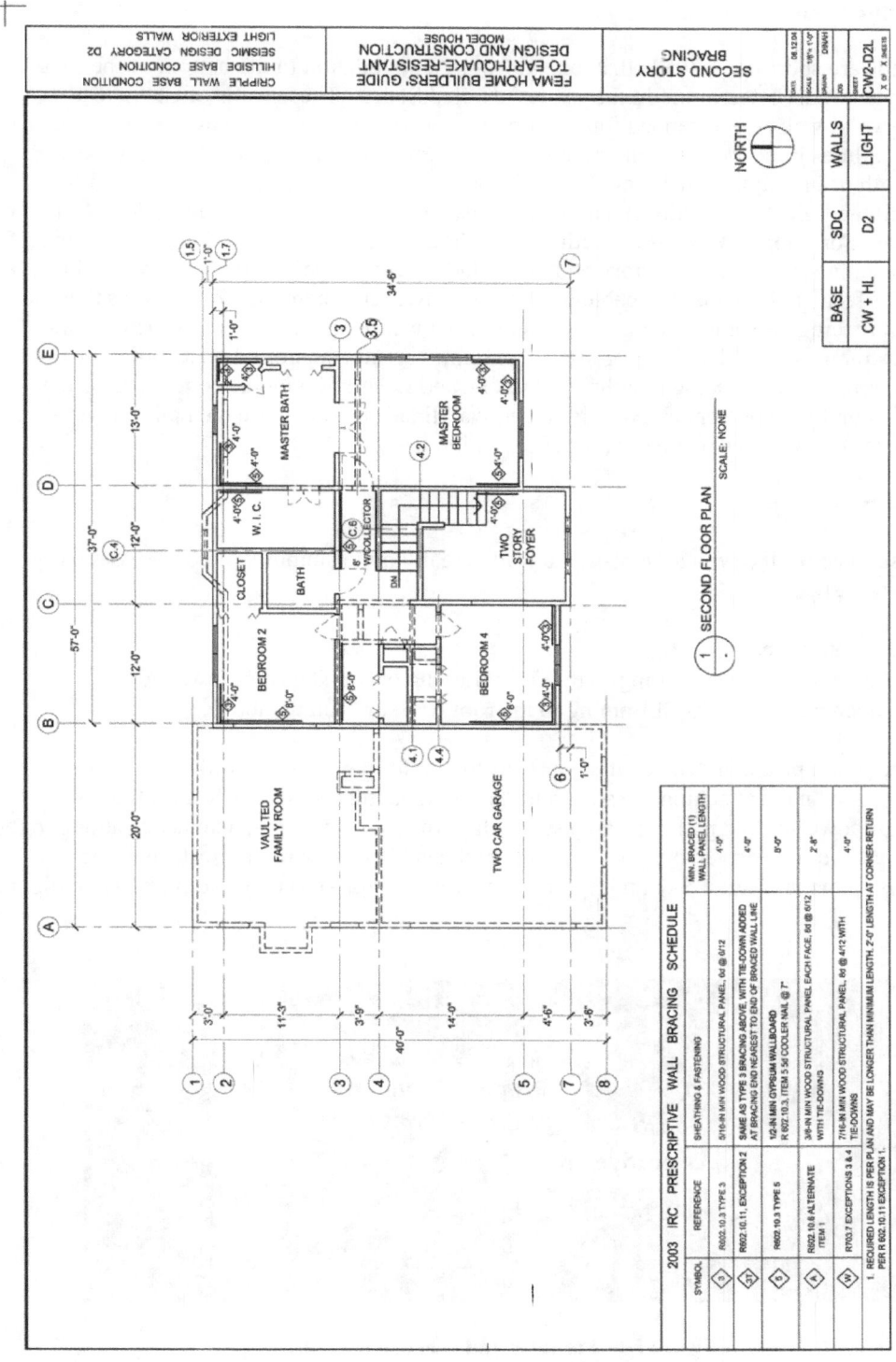

Figure 5-7c Plan view of second floor for model house with light-weight finish material located in SDC D_2.

5.1.3 Cripple Walls

Cripple walls are short frame walls that extend from the foundation to the bottom of the first floor. They are most often found in the western United States. These walls often enclose a crawl space or serve as walls for a stepped foundation. Historically, these walls have been the cause of significant failures in residential construction during earthquakes primarily due to inadequate in-plane strength or inadequate anchorage to the foundation. These walls are the most highly loaded of all the light-frame walls in a house because they have to resist the entire load from the house above. For cripple walls, *IRC* Section R602.10.2.1 specifies the length of bracing as 1.15 times the bracing required for the story above and indicates that the spacing should be 18 feet instead of 25 feet. This is not applicable in SDC D_2, however, where the *IRC* requires that the cripple wall bracing be a minimum of 75 percent of the wall length and be constructed using wood structural panels. When interior braced wall lines are not supported on a continuous foundation, cripple wall bracing lengths in SDCs D_1 and D_2 must be modified to increase the sheathing length by 50 percent at exterior braced wall lines or to decrease the nail spacing along sheathing edges to 4 inches on center (per *IRC* Section R602.10.11.1).

5.1.4 Quality Assurance

Quality assurance for the typical light-frame wall is really quite simple. There are essentially three areas to inspect:

- The sheathing nails,
- The anchorage of the framing to the floor framing or foundation below, and
- The anchorage of the wall framing to the roof or floor framing above.

The most common problem that adversely affects the performance of all wall types is the overdriving of the nails attaching the sheathing to the studs; this is especially a problem when pneumatic or power-driven nail guns are used. All of the nails used to attach the sheathing to the wall framing should be driven only to where the nail head is flush with the surface of the sheathing as shown in Figure 5-8; an improperly driven (overdriven) nail also is shown in Figure 5-8.

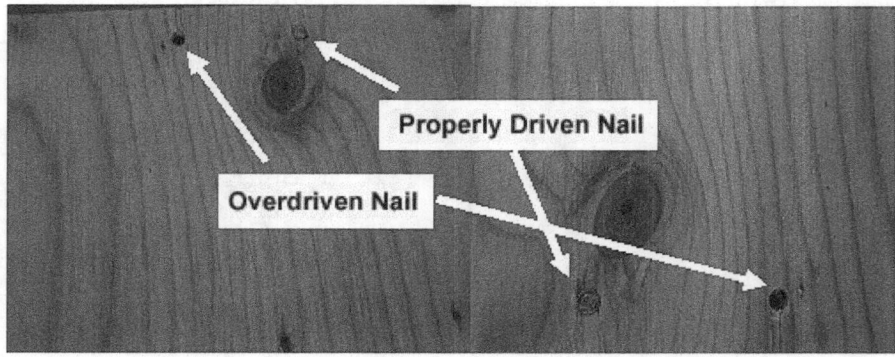

Figure 5-8 Properly and overdriven nails.

The most important nails are those around the perimeter of each panel of sheathing. These nails govern the strength and stiffness of the panel. If the nails are overdriven (as in Figure 5-8), the strength of the connection is severely compromised. For instance, if 3/8-inch wood structural panel sheathing is used and the nails are overdriven 1/8 inch (typical for many pneumatic nail tools), the strength of the wall is reduced as much as 40 to 50 percent. It is therefore imperative that the sheathing nails be inspected to ensure that they are properly driven. Many pneumatic nail tools have default driving pins that overdrive the nails but changing the driving pin costs little (less than 10 percent of the original cost of the tool). Many tool manufacturers now provide an adjustment on the tool to allow the user to change the depth of the driving pin without replacing the part. (It is recommended that where nail heads occasionally are more than 1/16-inch below the surface, an additional nail should be provided between existing nails. If a substantial number of nails are overdriven, the sheathing should be removed and the framing checked for splitting before replacing the sheathing with proper nailing.)

Nails often are located too close to the edge of the sheathing panel, which can result in the nails tearing out the side of the panel and weakening the wall. Therefore, the minimum edge distance for nailing sheathing is 3/8 inch. The larger the edge distance is, the stronger the wall will be. This is especially true for the row of nails at the bottom of the wall. If the bottom row of nails is located at the mid-height of the bottom plate for the wall, the displacement capacity of the wall is almost double that when the nails are spaced at 3/8 inch. When the edges of two sheathing panels meet on a common stud (2x nominal), the maximum edge distance should be 3/8 inch to prevent the nail from splitting the edge of the stud behind the sheathing.

Another common problem is that the nails are driven through the sheathing but miss the stud behind. Although it is obvious that nails in the air provide no strength or stiffness, this problem is so common it merits emphasizing that sheathing nails must be driven into the studs and top and bottom plates of the wall and not miss the framing. The use of pneumatic nail guns makes it very difficult for the operator to "feel" whether or not the nail has been driven into the stud behind; therefore, all such nailing must be visually inspected from behind to ensure that the nails did not miss the studs.

Finally, the sheathing nailing around the perimeter of each sheet of sheathing should be symmetric about the center of the sheathing panel. This means that there should be approximately the same number of nails along each parallel side of the panel as illustrated in Figure 5-9.

The second area to inspect for quality assurance is the anchorage at the bottom of the wall. For prescriptive walls, the nails must pass through the bottom plate into the floor framing. If the nails miss the floor joists, they have no capacity to resist withdrawal and sliding. If the wall is attached directly to the foundation, 1/2-inch-diameter anchor bolts should be spaced at a maximum of 6 feet on center, and the end bolts should be not more than 12 inches nor closer than 7 bolt diameters from the end of the plate. If the house is located in SDC D_1 or D_2, the anchor bolts are required to have a steel plate washer between the wood sill plate and the nut. (See Section 3.10 of this guide for more discussion of the use of plate washers.) If the walls are segmented (engineered), the hold-down connector needs to be attached to the end stud (usually a double stud) and the bolt or threaded rod holding the connector to the foundation or story below needs to be tightened to a snug one-quarter turn condition. Tightening the nut more does not provide higher reliability due to the fact that wood is viscoelastic and the tension in the bolt will not be sustained over time. A common practice to prevent the nuts from loosening during an earthquake is to use a double nut or slack take-up device. The hold-down should connect the end stud either directly to the foundation or to the top of a stud in a wall of the story below in order to effectively transmit the overturning loads into an active resisting element.

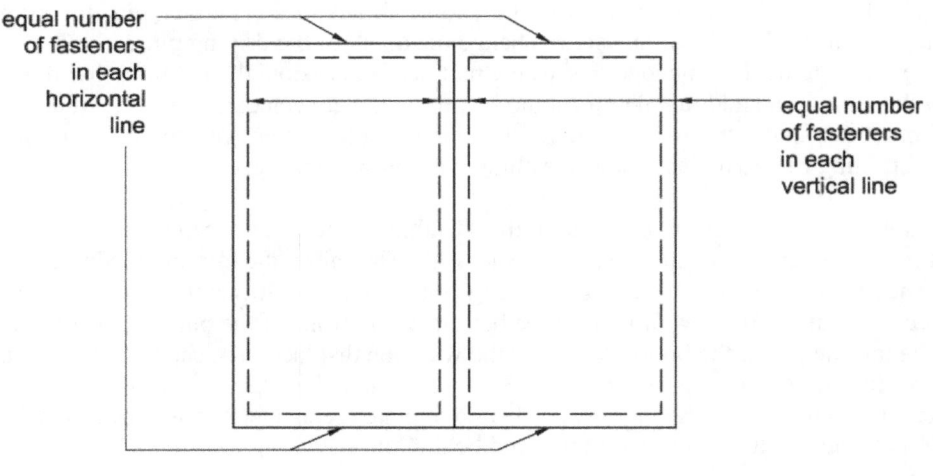

Figure 5-9 Illustration of concept of equal numbers of fasteners in line for symmetric nailing schedule.

Finally, the connection between the top plate and the roof framing or floor framing for the story above should be checked. The lateral loads are transmitted through nails or other fasteners connecting the top plate of the wall to the floor or ceiling joists above. (For vaulted ceilings, the connection might be between the top plate and the rafters.) If there is a wall on the story above that has an overturning anchor attached to the end stud, there should be an equivalent connection near the top of a stud in the story below that is upside down from the one on the wall above. A threaded rod should connect the two hold-down anchors to transmit the loads between them. If a stud is being used to transmit overturning loads from the story above, it will need to have an overturning anchor at its base to transmit the loads to the next story or foundation below.

Above-code Recommendations: Improving the detailing of the braced wall system is the most effective way to obtain earthquake performance levels higher than the code minimums. It is also most effective to concentrate on the lower stories of the house since this is the area that typically has the fewest walls and experiences the highest loads during an earthquake.

Use of continuous structural panel sheathing is recommended as an above-code measure (see Section 5.1.1 and Figure 5-5b of this guide). The analyses of the model house used in this guide indicated that use of continuous wood structural panel sheathing with overturning anchors at corners significantly reduced the drift in all SDCs and improved the performance category from "significant damage" to "moderate damage" in SDC D_2. The cost of making this change would be approximately 9 to 10 percent of the cost of the structural portion of the model house. The percentage of the total cost of the house would depend on the level of finishes, fixtures, windows, etc., and the cost of the land relative to the cost of the structural portion of the house. The structural portion of house costs ranges from 15 to 25 percent of total cost; therefore, the cost of fully sheathing a typical house would be approximately 1.5 to 2.5 percent of its total cost.

Use of wood structural panel sheathing extending over and nailed to the floor rim joist or blocking is recommended as an above-code measure. This is illustrated in Figure 5-10 and can be accomplished either by sheathing the wall with oversized panels (9-foot panels on an 8-foot wall) or by cutting and blocking standard size sheets. It is important to leave a vertical gap between sheathing panels at the mid-height of the band joist (approximately 3/4 inch for green solid sawn floor framing) to allow for shrinkage of the wood floor member without causing the sheathing to buckle. Analytical studies of the model house used in this guide indicated that use of sheathing spliced on the rim joist increased the approximate performance category from "significant damage" to "moderate damage" in SDC D_2. Improvements identified in the analytical study were less significant in SDCs C and D_1. For the model house, the cost of implementing this improvement would be approximately 0.5 percent of the cost of the structural portion of the house, essentially a no-cost item when the total cost of the house is considered.

Use of hold-down anchors at each end of each wood structural panel wall segment is recommended as an above-code measure (see Section 5.1.1 and Figure 5-5c of this guide). Analytical studies of the model house used in this guide indicated that the addition of hold-down anchors significantly reduced drift in all SDCs and the approximate performance category was increased from "significant damage" to "moderate damage" in SDC D_2. For the model house, the cost of this improvement would be 18 percent of the structural cost of the house or 2.5 to 4.5 percent of its total cost. This option provides a house with the best connectivity and is the basis for engineered shear wall construction. Further, of all the options, it has the greatest effect on the stiffness and strength of the wall.

Above-code Recommendation: **Although not addressed in the analytical studies, use of the above-code measures in combination is thought to have a cumulative effect and is recommended.** This level of connectivity will further improve earthquake performance by stiffening and strengthening the walls. **Additional above-code options for increasing strength and stiffness include spacing sheathing nails closer than the standard 6 inches on center (additional detailing requirements may be applicable at 3-inch or closer spacing; see the *IBC*) and placing wood structural panel sheathing on both faces of walls.** Use of these **above-code** recommendations will be most effective where the highest loads are present, such as lower stories and cripple wall levels.

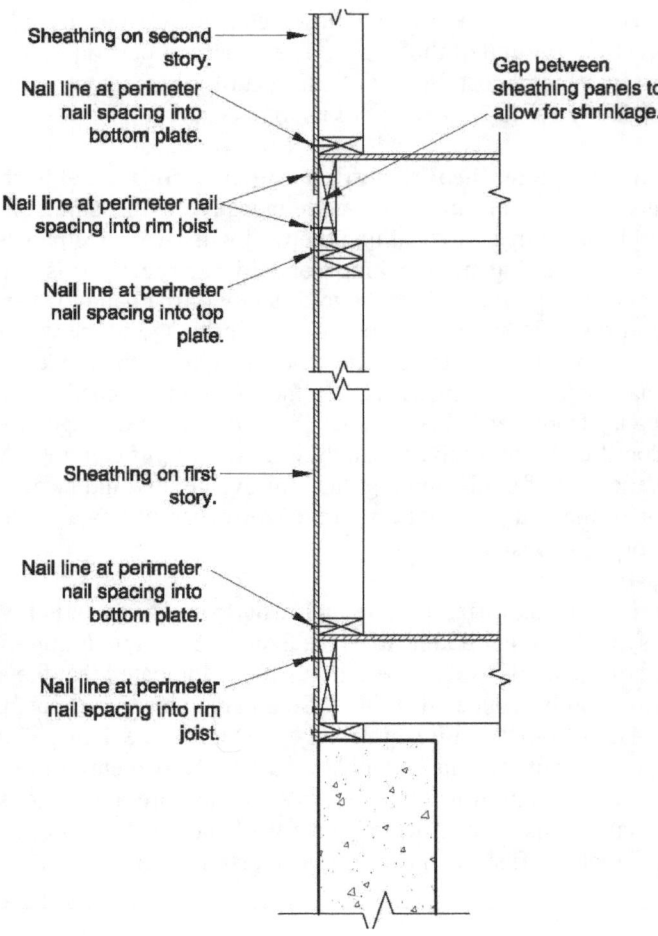

Figure 5-10 Sheathing detail for extending the sheathing over the band joist.

5.2 STONE AND MASONRY VENEER

Masonry and stone veneer are popular exterior finish materials for residential construction (Figure 5-11). Veneer provides a durable finish for the house but, unfortunately, the added weight of these veneers increases the loads experienced during an earthquake. The *IRC* permits use of stone and masonry veneer installed over concrete or masonry walls and over a backing of wood light-frame or cold-formed steel construction.

Figure 5-11 Applications of masonry and stone veneer.

This guide section discusses general principles of earthquake-resistant design for houses with veneer, specific *IRC* requirements important to earthquake performance, and **above-code** measures for improved earthquake performance.

Stone and masonry veneer is addressed in *IRC* Sections R702.1 (interior) and R703.7 (general and exterior). For earthquake-resistant construction using veneer, there are two major areas of concern:

- The increased earthquake loading on the house due to the weight of the veneer and
- Adequate anchorage of the veneer.

The weight of stone and masonry veneer permitted under the *IRC* provisions can vary from as little as 20 pounds per square foot installed for adhered veneer to as much as 70 psf installed for a 5-inch-thick anchored veneer with 1 inch of grout. The weight of stone and masonry veneer greatly increases the overall weight of a light-frame house and, as a result, the earthquake loads. The *IRC* provisions rely exclusively on the strength and stiffness of the light-frame bracing systems to resist wind and earthquake loads, discounting any strength and stiffness that might be provided by the veneer. This is primarily because there is only limited understanding of the ability of the veneer to resist cyclic earthquake loading while acting in combination with the light-frame bracing systems and because the veneer cracks and breaks at smaller displacements than required for the light-frame system to achieve its capacity. As a result, additional requirements apply to light-frame systems with veneer in areas with higher earthquake risk.

Masonry and stone veneers have been damaged in historic earthquakes. Figure 5-12 illustrates damage that occurred during the Northridge earthquake that appears to be due to inadequate connection of the wall studs to the top and bottom plates. In failing, the stone veneer pulled the wall sheathing and many of the studs out of the wall perpendicular to the pictured wall.

Figure 5-12 Stone veneer damaged during the earthquake in Northridge, California.

5.2.1 *IRC* Earthquake Requirements

The *IRC* permits use of veneer above the first story above grade in SDCs D_1 and D_2 only for one- and two-family detached houses of wood light-frame construction without cripple walls.

If a house does not meet these requirements, a design professional should design the house following the requirements of the *International Building Code* or *NFPA 5000*. For veneer installed over a light-frame system, requirements are incrementally more restrictive for higher SDCs (*IRC* Section R703.7). References to the veneer provisions for SDCs C, D_1, and D_2 appear in *IRC* Sections R301.2.2.3.1 and R301.2.2.4.2.

For light-frame systems, the general rule is that stone and masonry veneer not exceeding 5 inches in thickness is permitted in the first story above grade across all SDCs without any further requirements (*IRC* Section R703.7).

Although this veneer will have some impact on earthquake performance, the increase in earthquake load close to the house base is not judged to have significant life-safety implications.

In SDCs A and B, veneer on wood light-frame or cold-formed steel houses can be up to 5 inches thick and extend up to 30 feet above a noncombustible foundation with an additional 8 feet permitted for gable end walls. The 30-foot limit corresponds to the height at which additional vertical support of the veneer would be required in an engineered design. Veneer in SDCs A and B has no imposed earthquake limitations, leaving wind provisions to control.

In SDC C, veneer on wood light-frame or cold-formed steel houses can be up to 5 inches thick and extend up to 30 feet above a noncombustible foundation with an additional 8 feet permitted for gable end walls. It is required, however, that the length of bracing walls in other than the topmost story be 1.5 times the length required if veneer is not used. (Note that this provision is applicable to all houses in SDC C.)

In SDC D_1, wood light-frame houses are permitted to have veneer up to 4 inches in thickness for a height of 20 feet above a noncombustible foundation with an additional 8 feet allowed for gable end walls or up to 30 feet where the bottom 10 feet (first story) is anchored to a concrete or masonry wall. However, *IRC* Section R301.2.2.2.1 assembly weight limits for concrete and masonry leave little allowance for veneer weight. Where assembly weight limits cannot be met, engineered design of the concrete or masonry walls may be required.

Veneer on wood light-frame walls in SDC D_1 (see *IRC* Section R703.7) is dependent on the use of wood structural panel sheathing with specified nail size and spacing, increased bracing lengths, use of hold-downs with specified capacities at each end of each braced wall panel, and the prohibition of cripple walls as previously mentioned. The minimum sheathing must be at least 7/16 inch thick and it must be fastened with at least 0.131 x 2.5 inch (8d common) nails that are spaced no more than 4 inches on center at the panel edges. The braced wall length for the top story of the house must be at least 45 percent of the total wall length. Finally, the first two stories must have hold-down connectors installed. The hold-down connectors for the transfer of loads from the second to the first story must have a minimum capacity of 2100 lb and from the first floor to foundation must have a minimum capacity of 3700 lb. Load path requirements with hold-down devices are discussed in Chapter 2 of this guide.

In SDC D_2, veneer provisions are similar to those for SDC D_1, except for a couple of small changes:

- The maximum thickness of the veneer is limited to 3 inches,

- The top story must be sheathed to 55 percent of the total wall length,

- The second-to-first story overturning anchors must have a capacity of 2300 lb, and

- The first-story-to-foundation overturning anchors must have a minimum capacity of 3900 lb.

Anchorage requirements for anchored stone or masonry veneer are given in *IRC* Section R703.7.4. Use of galvanized corrugated sheet metal ties or metal strand wire ties is required. The minimum gage required is No. 9 U.S. gage wire for strand wire ties and No. 22 U.S. gage by 7/8 inch corrugated sheet metal ties. The maximum supported veneer area is 2-2/3 square feet per tie for SDCs A, B, and C. The supported area is reduced to 2 square feet per tie for houses in SDC D_1 and D_2.

The intent of using ties for masonry veneer is not to prevent the veneer from cracking but rather to prevent the veneer from pulling away from the supporting wall system. In earthquakes, the desire is to prevent damaged veneer from becoming a falling hazard and the use of ties should accomplish this by holding the cracked veneer to the supporting wall.

> **Above-code Recommendations: The sheet metal ties or wires used to fasten veneers should be corrosion resistant, should penetrate the house paper and sheathing, and should be embedded into the wall studs.** Fasteners that do not penetrate the studs (i.e., only the sheathing is penetrated) have low withdrawal resistance and significantly reduce the tie's ability to hold the veneer to the wall.
>
> **Where veneer is limited to the first story above grade, increase the length of wood structural panel bracing and use hold-down devices on the braced wall panels in the first story to increase both the strength and stiffness of the first story above grade.** These measures will help make the deformation behavior of the light-frame system more compatible with the veneer and reduce cracking of the veneer.

5.2.2 Quality Control

All mortar joints should be completely filled and well tooled for water tightness. This can have a significant effect on the strength and durability of the masonry walls and veneers. To augment the tooling of joints for improving moisture control (which directly affects the longer-term strength of the wall), flashing and weep holes need to be placed at the bottom of the veneer so that any water trapped between the veneer and the structural wall behind can be directed to the outside of the house. The flashing and weep holes frequently are omitted in residential construction, resulting in expensive repairs to the structural walls of the house.

Veneer should be placed such that a 1-inch cavity is maintained between the veneer and the supporting wall. This cavity provides a drainage plane for channeling the moisture that will migrate from the outside surface of the veneer to the inside and then to the bottom of the wall where it can escape to the outside through the weep holes. This drainage plane is very important to prevent the wall sheathing from being constantly wet, which will result in mold, mildew, and decay.

Anchor ties that hold the veneer to the wall need to be placed at the proper spacing to ensure that the area of veneer attributed to each tie does not exceed the maximum allowed for the particular Seismic Design Category. For SDCs D_1 and D_2, the spacing should be such that each tie is supporting no more than 2 square feet of veneer. The ties also need to be placed such that the nail holding the tie to the wall is embedded into a stud and not just the sheathing material.

Finally, if horizontal reinforcing wire is used, it should be a minimum size of W1.7 and should be placed at a maximum vertical spacing of 18 inches according to the 2002 *Masonry Standards Joint Committee Code* (ACI 530/ASCE 5/TMS 402, Section 6.2.2.10.2.4.). This wire is embedded in the horizontal mortar joints between the courses of brick. The brick ties that hold the veneer onto the wall behind it should be bent around the wire to improve the effectiveness of the tie in preventing the veneer from falling off the wall during an earthquake.

5.3 COLD-FORMED STEEL HOUSES

The *IRC* permits cold-formed steel framing prescriptive construction. This guide section discusses the general principles of earthquake-resistant design for cold-formed steel houses, specific *IRC* requirements important to earthquake performance, and **above-code** measures for improved earthquake performance.

Cold-formed steel construction is, in most respects, like wood light-frame construction. The *IRC* contains provisions similar to the wood light-frame provisions for cold-formed steel floors, roofs, and walls. The weights of cold-formed steel floor, roof, and wall assemblies are the same or slightly lower than those with wood light-frame construction and result in very similar earthquake loads. The system resisting wind and earthquake loads most commonly consists of floor and roof assemblies acting as horizontal beams carrying loads to shear walls. Cold-formed steel shear walls carry wind and earthquake loads to the foundation. The *IRC* provides extensive illustrations of minimum construction requirements for cold-formed steel framing. One significant difference between steel and wood light-frame construction is that steel construction assumes that the framing is in-line – that is, that the rafters, studs, and floor joists line up so that the top channels of the walls and other elements are not subjected to compression and bending loads. Figure 5-13 shows a cold-formed steel house under construction.

Figure 5-13 Cold-formed steel house under construction.
Photo Courtesy of Dietrich Metal Framing -- a Worthington Industries Company

5.3.1 Scope Limitations

The *IRC* limits the scope of cold-formed steel houses permitted under the prescriptive provisions. Some scope limitations are found in *IRC* Chapter 3 while others are found in *IRC* Section R603.

For all SDCs, steel light-frame houses are limited to two stories above grade; the maximum permitted plan dimensions are 60 feet perpendicular to truss or joist span and 36 feet parallel to truss or joist span (*IRC* Sections R505.1.1, R603.1.1, R804.1.1). For all SDCs, story height is limited to a 10 foot stud clear height plus a height of floor framing not to exceed 16 inches (*IRC* Section R301.3). For houses in SDCs D_1 and D_2, *IRC* Section R301.2.2.2.1 limits cold-formed steel wall assembly weights to 14 psf for exterior walls and 5 psf for interior walls. Floor and roof plus ceiling assembly weights are also limited to 15 psf by *IRC* Section R301.2.2.2.1.

Although some interpretations of the *IRC* assume the code does not directly require that the irregularity limits of *IRC* Section R301.2.2.2.2 be followed, the same limits are imposed by AISI's *Standard for Cold-formed Steel Framing – Prescriptive Method for One- and Two-Family Dwellings* (2001) for SDCs D_1 and D_2. Attention to the issues associated with irregularities is important to prevent torsional response overloading of critical elements in the wall.

As for wood houses, the design of cold-formed steel houses is typically controlled by wind rather than earthquake loading through SDC C; as a result, there are no added *IRC* earthquake load requirements for SDC C. Per *IRC* Section R301.2.2.4.5, houses in SDCs D_1 and D_2 must conform to the requirements of the 2001 AISI *Standard for Cold-Formed Steel Framing – Prescriptive Method for One- and Two-Family Dwellings* (AISI/COFS/PM).

5.3.2 *IRC* Wind and Earthquake Requirements

Unlike wood light-frame systems, the cold-formed steel provisions rely exclusively on bracing at exterior walls for wind and earthquake load resistance. *IRC* Section R603.7 requires that the exterior walls be fully sheathed. *IRC* Table R603.7 specifies minimum percentages of full-height sheathing as a function of wind speed and exposure, roof slope, and story being braced. Per *IRC* Section R603.7, the minimum braced length is not permitted to be less than 20 percent of braced wall length and bracing panels must be a minimum of 4 feet in length to be counted towards the required percent with a minimum 4-foot-long bracing panel required at each corner of each exterior wall. The minimum wood structural panel sheathing thickness is specified as 7/16 inch for oriented strand board and 15/32 inch for plywood. The minimum edge fastening is spaced 6 inches on center using No. 8 screws. All wood structural panel edges are to be blocked.

A number of modifying factors are required to be applied to the tabulated percentage of full-height sheathing. The tabulated bracing percentages are based on walls with an 8-foot clear height and must be multiplied by 1.10 for a 9-foot clear height and 1.20 for a 10-foot clear height (*IRC* Section R603.7). The required sheathing percentage is permitted to be multiplied by 0.6 where hold-down devices having 4,300 lb capacity are provided at each end of each exterior wall (*IRC* Section R603.7.2). The minimum percentages of sheathing are based on a house aspect ratio (ratio of greater plan dimension to lesser plan dimension) of 1:1. The minimum percentages for the shorter walls must be multiplied by 1.5 and 2.0 for house aspect ratios of 1.5:1 and 2:1 respectively (*IRC* Table R603.7, footnote c). The minimum percentages are permitted to be multiplied by 0.95 or 0.90 for hipped roofs, depending on roof slope (*IRC* Table R603.7, Footnote d).

AISI/COFS/PM provisions for bracing in SDCs D_1 and D_2 are significantly different from *IRC* requirements for areas of less earthquake risk with respect to bracing approach, percentage of full-height sheathing, and detailing. The *IRC* basic bracing provisions (SDCs A, B, and C) do not require use of hold-downs and permit reduction of the percentage of full-height sheathing if hold-downs are provided. In contrast, AISI/COFS/PM basic bracing provisions for SDCs D_1 and D_2 require use of hold-downs at each end of each required full-height sheathing segment. An alternative approach allows hold-downs to only occur at each end of a braced wall line provided that the length of full-height sheathing is increased (up to 3x length otherwise required). This alternative approach is similar to the continuous structural panel sheathing (perforated shear wall) method used in wood light-frame construction. The user is referred to the AISI/COFS/PM document for full details of bracing requirements.

IRC Section R505 addresses floor construction using cold-formed steel framing. The use of wood structural panel floor sheathing is implied by details and fastening requirements. *IRC* Section R503 controls selection of floor sheathing depending on the spacing of joists and material used. *IRC* Table R505.3.1(2) requires a minimum fastening of No. 8 screws at 6 inches on center on the perimeter of the sheathing panels and 10 inches on center in the field. This is assumed to be at supported edges with no requirements for blocking at other edges. Similarly, *IRC* Section R804 implies wood structural panel roof framing, while *IRC* Table R804.3 requires minimum No. 8 screws at 6 inches on center at edges, 12 inches on center in the field, and 6 inches on center at gable end trusses.

It is very important to use appropriate framing materials and fasteners for the performance of the wall bracing and floor and roof systems to be acceptable. *IRC* Sections R505.2, R603.2, and R804.2 specify material grade, corrosion protection, and identification of the materials required. *IRC* Sections R505.2.4, R603.2.4, and R804.2.4 specify fastener requirements. Steel to steel and wood structural panel sheathing to steel connections are required to be made with self-drilling tapping screws conforming to SAE J78 with a Type II coating in accordance with ASTM B633. Screws are required to extend through the steel a minimum of three exposed threads. Steel to steel connections are to be made with self-drilling tapping screws at a minimum edge and center-to-center distances of 0.5 inches. Connections are typically specified with No. 8 screws, but adjustments are given for use of No. 10 and No. 12 screws. Wood structural panel sheathing to steel framing connections are to be made with No. 8 self-drilling tapping screws at minimum head diameter of 0.292 inches, countersunk heads, and a minimum edge distance of 3/8-inches. Gypsum wallboard ceilings are to be attached with minimum No. 6 screws conforming to ASTM C 954.

> **Above-code Recommendations:** There are no *IRC* maximum spacing limitations between lines of cold-formed steel bracing walls corresponding to the wood light-frame braced wall line spacing limits of 25 and 35 feet. With wood light-frame construction, these spacing limitations are intended to help distribute the bracing walls in proportion to the house mass (dead load) as well as to limit the loading to the floor and roof. This helps to reduce concerns regarding house rotation due to torsional irregularities and concerns regarding irregular floor and roof system shapes. For cold-formed steel houses, the maximum 60-foot house dimension provides some limit for earthquake loads in the floor and roof; however, **for improved earthquake performance, it is recommended that interior cold-formed steel braced wall lines be added such that the distance between braced wall lines does not exceed 35 feet**. This will help lower the loads within any given wall segment and better distribute the earthquake loads throughout the house. The cost associated with adding additional bracing walls would be consistent with the percentage increase in wall length.
>
> **For cold-formed steel houses in all SDCs, the irregularity limitations developed for wood light-frame houses should be applied.**

5.3.3 Quality Control

Quality control for cold-formed steel construction can play a major role in the satisfactory performance of the house. Important steps in steel construction include ensuring proper load path connections. Just as in wood light-frame construction, the connections between the end studs and the floor or foundation become critical for the strength and stiffness to be the highest possible. Strong connections ensure the load path will function properly.

Adequate lateral bracing of all floor, roof, and wall framing members is important to prevent the individual members (joists or studs) from buckling. Cold-formed members have to be braced well before they can support significant compression loads. Attaching sheathing to the stud or joist typically will provide this lateral support.

Screws that attach the sheathing to the framing should be inspected to ensure that they are not overdriven (the resulting loss in capacity will be similar to that experienced in wood light-frame construction when nails are overdriven). Just as in wood framing, the power tools used to install the screws can easily over drill the screws, leaving the head of the screw below the surface of the sheathing material and significantly increasing the chances of pull through.

5.4 MASONRY WALL HOUSES

While many houses use masonry walls for foundations, some houses are built using masonry for the walls above grade. Figure 5-14 shows a house constructed using masonry for the walls. The *IRC* permits construction of houses with masonry walls in accordance with prescriptive provisions for concrete, brick, and stone masonry. General provisions for masonry construction appear in *IRC* Section R606. Additional details of construction for concrete unit masonry, brick masonry, and masonry grouting appear in *IRC* Sections R607 through R609. However, it is important that all masonry cells with reinforcement be grouted, and care should be taken to ensure that the grout is consolidated. Discussed below are the general principles of earthquake-resistant design for masonry wall houses, specific *IRC* requirements important to earthquake performance, and **above-code** measures for improved earthquake performance.

Figure 5-14 **House constructed with masonry walls.**

Masonry wall construction is significantly heavier than light-frame wall construction; however, in most respects, the same principles of earthquake-resistant construction apply. The system resisting wind and earthquake loads most commonly consists of floor and roof assemblies acting as horizontal beams carrying loads to bracing walls. Masonry bracing walls carry wind and earthquake loads down to the foundation with masonry walls primarily resisting loads acting in their strong direction – that is, parallel to the wall. Because masonry walls are heavier than stud framing, it follows that the earthquake loads resisted by each wall element and connection will be greater. As a result, the design of the connections is as important as the proportioning of the floor and roof systems and bracing walls. Significant earthquake loads also develop perpendicular to the wall surface due to the wall's weight. These loads tend to pull the walls away or push them toward the floor or roof, making wall to floor or roof anchorage critical. Damage from insufficient anchorage or connectivity similar to that shown in Figure 5-12 can result if these connections are not strong enough.

5.4.1 Scope Limitations

One of the primary approaches the *IRC* uses to deal with the increased earthquake loading associated with masonry wall houses is to limit the scope of houses permitted under the prescriptive provisions. *IRC* Section R301.2.2 introduces earthquake provisions applicable to houses in SDCs D_1 and D_2. The first scope limitation is found in *IRC* Section R301.2.2.2.1, Weights of Materials, which limits masonry walls to 8 inches thick and 80 pounds per square foot, which essentially prohibits the use of rubble stone masonry (*IRC* Section R606.2.2) as well as relatively thick concrete or brick masonry walls. *IRC* Section R301.2.2.4.3 triggers the reinforcement and configuration requirements of *IRC* Section R606.11.3 for all houses in SDC D_1 and of Section R606.11.4 for all houses in SDC D_2. In SDCs D_1 and D_2, the scope of prescriptive construction of houses with masonry walls is effectively limited to walls one story in height with up to 9 feet between lateral supports; beyond this, engineered design is required. A final scope limitation in *IRC* Section R301.3, Item 3, limits masonry house walls to a clear height of 12 feet with a maximum floor framing depth of 16 inches. An additional 8 feet of masonry wall height is permitted for gable end walls. It is also worth noting that *IRC* Section R301.2.2.2.2, Item 7, prohibits the mixing of light-frame and masonry construction such that light-frame walls would be required to support earthquake loads due to masonry wall construction (e.g., mixing masonry and light-frame walls on the same story.)

5.4.2 Wall Lateral Support, Reinforcing, and Anchorage

General requirements for lateral support of masonry walls for all SDCs are provided in *IRC* Section R606.8. Walls are permitted to be laterally supported by cross-walls, pilasters, buttresses, or structural frame members where walls span horizontally between supports and by floors and roofs where walls span vertically between supports. Minimum reinforcing requirements apply only for masonry laid in stack rather than running bond and for interior non-load bearing walls where they intersect with other masonry walls. *IRC* Section R606.10 requires anchorage of masonry walls to floor and roof systems in accordance with specific details. However, in recent U.S. earthquakes, similar details have been observed to be susceptible to damage. Performance of some of these details for supporting site-built masonry walls also is thought to be problematic. Several details on how to comply with the *IRC* provisions are provided in the code itself.

IRC Section R606.11.1.1 triggers additional requirements for floor and roof systems for houses in SDCs D_1 and D_2. Floor and roof wood structural panel sheathing is required to have all edges blocked and nailed at 6 inches on center. Where the floor or roof system is long and narrow (more than twice as long as it is wide), the nailing at sheathing edges is required to be reduced from 6 to 4 inches on center.

For houses in SDC D_1, *IRC* Section R606.11.3 places further limits on masonry walls using prescriptive provisions:

- Masonry walls are limited to one-story,
- Additional reinforcing is required,
- Reinforcement detailing provisions for masonry columns apply, and
- Type N mortar and masonry cement are prohibited.

Similar additional requirements are triggered by *IRC* Section R606.11.4 for SDC D_2. SDCs D_1 and D_2 requirements are illustrated in *IRC* Figure R606.10(3). Again, specific design is required for wall to framing anchorage per *IRC* Section R606.11.2.2.1; thus, the anchorage to roof framing illustrated in *IRC* Figure R606.10(3) may not be the only alternative.

In the design of wall anchorage to the floor and roof, it is important that a direct tension tie be provided from the wall to the floor or roof framing members if framing is perpendicular or to blocking that is continued across the floor or roof if framing is parallel. Detailing in *IRC* Figures

611.8(1) to (7) shows the concept of this connection when used with concrete walls, but they are equally applicable to masonry construction.

5.4.3 Parapet walls

IRC Section R606.2.4 limits masonry thickness and height of parapets in all SDCs and sets the requirements for reinforcing walls of houses in SDCs D_1 and D_2. Parapet walls are easily damaged in an earthquake and attention to placement of reinforcement and bracing to stabilize the parapet is important.

5.4.4 Problematic Gaps Between Prescriptive and Engineered Construction

The current *IRC* requirements for masonry construction reveal significant gaps between prescriptive and engineered construction, creating a significant opportunity to employ **above-code** measures to improve performance.

For areas of high earthquake risk, one notable omission is the lack of a limit on the amount of opening in masonry walls. Light-frame provisions require a percentage of the wall to be solid to provide bracing so that the bracing wall strength is in proportion to the earthquake load, but the same requirement is not imposed for masonry walls. This is of particular concern when the house layout results in front and back walls with a large percentage of openings for doors and windows.

> **Above-code Recommendation: Each exterior wall and each interior bracing wall should have at least one, and preferably two, sections of solid wall not less than 4 feet in length. Further, sections of solid wall should not be spaced more than 40 feet on center and should be placed as symmetrically as possible** (Figure 5-15). The provisions of *IRC* Section R611.7.4 for ICF walls illustrate a reasonable approach to regulation of minimum wall length. The improvement requires only reasonable planning and should not result in higher construction costs.

IRC Section R403.1 requires that exterior walls be supported on continuous concrete or masonry footings and regulates minimum footing width and depth, but nothing in the *IRC* appears to require that other masonry walls be supported on foundations.

> **Above-code Recommendation: It is vital that all masonry walls be supported on substantial continuous footings extended to a depth that provides competent bearing**. If this is not done the walls have a high probability of being damaged due to uneven settlement.

There are no *IRC* maximum spacing limitations between lines of masonry bracing walls corresponding to the light-frame limits of 25 and 35 feet. In light-frame construction, these spacing limitations are intended to help distribute the bracing walls in proportion to the house mass (dead load) as well as to limit the loading to the floor and roof, which helps prevent damage to the house from rotation due to torsional irregularities caused by irregular floor and roof system shapes.

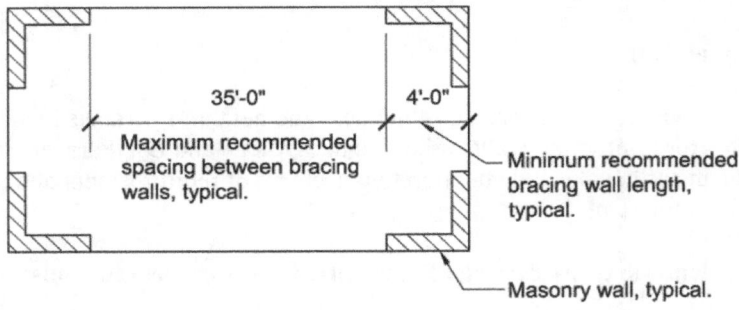

Figure 5-15 Symmetric layout of walls to distribute loads uniformly and thereby prevent torsion.

Above-code Recommendations: Because prescriptive detailing is not available for the addition of interior masonry braced wall lines, it is recommended that the distribution of bracing walls be carefully balanced and that the floor and roof plans use simple rectangular shapes without indentions, bump-outs, or openings. The cost associated with this improvement is proportional to the length of wall added in relation to the initial total wall length.

The *IRC* also does not regulate house irregularities for masonry construction. **The concepts behind *IRC* Section R301.2.2.2.2, Irregularities 1 through 6, should apply equally to a masonry wall house. The exceptions to Irregularities 2 and 5 also can be applied; however, the rest of the exceptions are not applicable.**

Above-code Recommendation: Solid portions of wall should be stacked from floor to floor and masonry walls should be continuous from the top of the structure to the foundation. Masonry walls not directly supported on walls below require engineered design for gravity load support. Design for earthquake and wind loads also should be provided. The cost of this improvement is negligible; it requires only proper planning when laying out the wall positions for the house.

Chapter 5, Walls

> **Above-code Recommendation: Running bond lay up of masonry units is inherently much stronger than a stack bond lay up. For concrete masonry, use of open end units at locations of vertical reinforcement and use of bond beam units for horizontal reinforcement also help to increase the interlocking of the masonry construction, thereby increasing strength.** The cost associated with using running bond rather than stacked bond masonry is minimal. The cost of bond beam masonry units for placing horizontal reinforcement at the top of each wall segment would be significantly less than 1 percent of the cost of the original structural system but would dramatically improve the performance of the wall and connection to the floor or roof framing.

> **Above-code Recommendation: Any of the many required or recommended measures for areas of high earthquake risk would improve the performance of masonry wall houses in areas of lower earthquake risk as well as in high-wind areas. Priorities include provision for reinforcing such as that shown in *IRC* Figure R606.10(2), wall anchorage using details developed to resist out-of-plane wall loads such as those shown in *IRC* Figures R611.8(1) to (7) for walls, minimum length of bracing walls, and a maximum spacing of bracing wall lines.**

5.4.5 Quality Control

Quality control during masonry construction can play a major role in the satisfactory performance of the masonry. The following should be monitored:

- Proper placement of reinforcing in masonry. Unless specifically designed otherwise, reinforcing should be located as near the centerline of the masonry cavity as possible. In no case should reinforcing be closer than 5/8 inch to a masonry unit wall.

- Anchor bolts should be secured into place before grout placement (not placed into grout following the pour).

- Excess mortar and other obstructions should be cleaned from the cavity to allow free placement of grout.

- Consolidation of the grout after it is placed into the cavity is necessary to eliminate voids.

5.5 INSULATING CONCRETE FORM (ICF) WALL HOUSES

The *IRC* permits construction of houses with concrete walls in accordance with the prescriptive provisions of *IRC* Section R611 for ICF (insulating concrete form) walls and Section R612 for conventionally formed walls. ICF walls are concrete that is cast into forms that remain in place to serve as house insulation. The code covers three geometries of ICF forms: flat, waffle-grid, and screen grid forms. Both the waffle and screen grid consist of concrete cast into a series of interconnected horizontal and vertical cores within the insulation. Interior and exterior wall finishes are applied over the ICF wall form. ICF walls may be used for the full height of the house or for the lower story only with light-frame walls above. The provisions of *IRC* Section R611 are based on the use of light-frame floor, roof, and ceiling assemblies in ICF wall houses. The light-frame construction can be either wood or steel light-frame; however, current detailing developed for areas of high earthquake demand provides solutions for wood light-frame floors and roofs. This section discusses general principles of earthquake-resistant design for ICF wall houses, specific *IRC* requirements important to earthquake performance, and **above-code** measures for improved earthquake performance. An example of ICF construction is shown in Figure 5-16.

Figure 5-16 Insulated concrete form house under construction.

ICF wall construction, like masonry, is significantly heavier than light-frame wall construction. In most respects, however, the same principles of earthquake-resistant construction apply. The system resisting wind and earthquake loads consists of floor and roof assemblies acting as horizontal beams carrying loads to bracing walls. ICF bracing walls carry wind and earthquake loads down to the foundation with ICF walls primarily resisting loads acting in their strong direction, parallel to the wall (shear walls). Because ICF walls are heavier than light-frame walls, the earthquake loads to be resisted by each wall element and connection will be greater. As a result, proportioning of the floor and roof systems and bracing walls is important as are the connections between the walls and the floor and roof framing. Significant earthquake loads also develop perpendicular to the wall surface due to the wall weight. These loads tend to pull the walls away or push them toward the floor or roof, making wall to floor and wall to roof anchorage very important.

5.5.1 Scope Limitations

One of the primary approaches the *IRC* uses to address the increased earthquake loading with ICF wall houses is to limit the scope of houses permitted under the prescriptive provisions. For ICF, some scope limitations are found in Chapter 3 while others are found in *IRC* Section R611. For all Seismic Design Categories, ICF wall houses are limited to two stories above grade. The maximum permitted house plan dimension is 60 feet, and framing clear spans are limited to 32 feet for floors and 40 feet for roofs (*IRC* Section R611.2). For houses in SDCs D_1 and D_2, *IRC* Section R301.2.2.2.1 limits the dead weight of concrete walls to 85 pounds per square foot for walls that are 6 inches of solid concrete. *IRC* Section R611.7.4 limits the minimum nominal thickness of the wall to 5-1/2 inches. Application of these criteria and the maximum unit weights in *IRC* Table R611.2 permits use of 5-1/2-inch flat ICF walls, 6- and 8-inch waffle-grid walls, and 6-inch screen grid walls. The weight of interior and exterior wall finishes is limited to 8 psf (*IRC* Section R611.2). Floor and roof plus ceiling assembly weights also are limited by *IRC* Section R301.2.2.2.1 as for the other materials.

For houses in SDCs D_1 and D_2, *IRC* Section 611.2 limits the scope to rectangular houses with a maximum floor and roof aspect ratio (length to width ratio) of 2:1 and requires that the ICF walls be aligned vertically (stacked) without cantilevers or setbacks. An additional limitation is that the gable end portion of walls must be of light-frame rather than ICF construction. Further, the top of the ICF wall must be anchored to a required attic floor as described in *IRC* Section R611.9.

For houses in SDCs D_1 and D_2, *IRC* Section R301.2.2.2.2 requires engineered design for irregular portions of houses. The irregularities are discussed in detail in an earlier section of this guide. The exceptions to irregularities are mostly applicable to wood-frame construction and so do not apply. The 2003 *IRC* as printed refers to irregularities in *IRC* Sections R301.2.2.7 and R301.2.2.9, which do not exist; the intent was to reference *IRC* Section R301.2.2.2. *IRC* Section R611.8.3.1 repeats the prohibition of vertical offsets in floor framing (such as split levels) in SDCs D_1 and D_2.

IRC Section R301.2.2.4.4 triggers the requirements of *IRC* Sections R611 and R612 for all houses in SDCs D_1 and D_2. These requirements include all of the reinforcement and detailing provisions for concrete walls that provide the continuity and connectivity for the system to function effectively under wind or earthquake loads.

5.5.2 Wall Reinforcing and Anchorage

Reinforcing steel is of primary importance for concrete and masonry house performance. The reinforcing steel is what ties the components together and provides toughness to the concrete when it cracks. General requirements for reinforcing of ICF walls are provided in *IRC* Sections R611.3 through R611.5. Reinforcing is required for all ICF walls across all SDCs. *IRC* Section R611.7.1.2 imposes additional minimum reinforcing size and spacing limits for houses in SDCs D_1 and D_2. *IRC* Section R611.7.1.3 provides similar limits for horizontal reinforcing and includes requirements to provide for continuity of reinforcing around house corners and termination of horizontal reinforcement. Reinforcement is required to be doweled into the foundation and extend for the full height of the ICF wall, including the parapet if applicable. Lap splicing of reinforcing is permitted.

Like light-frame wall bracing provisions, ICF wall provisions require a minimum length or percentage of the wall to be solid (without openings) so that the bracing wall strength is in proportion to the earthquake load. The minimum required lengths of bracing wall for ICF construction in each Seismic Design Category are defined in *IRC* Section R611.7.4. Minimum bracing lengths must be provided for wind load parallel and perpendicular to the ridge per *IRC* Tables 611.7(9A) through (10B). The required bracing length must be provided using wall segments not less than 2 feet in width. For all houses in SDCs D_1 and D_2, the bracing length requirements of *IRC* Table R611.7(11) also must be met using wall segments that are not less than 4 feet in width. The increase from 2 to 4 feet provides a wall that is stronger and much better able to withstand cyclic earthquake loading.

Also of primary importance for house performance is the connection of the ICF walls to the floors and roofs, both for bracing loads parallel to the wall and out-of-plane loads perpendicular to the wall. The *IRC* addresses three types of connections to wood light-frame floors and roofs:

- Top bearing connections where the light-frame floor bears on the top of the ICF wall with light-frame walls above (*IRC* Section R611.8.1),

- Ledger-bearing connections where the floor or roof is supported off the face of the ICF wall by a wood ledger (*IRC* Section R611.8.2), and

- Top-bearing connections where the light-frame roof bears on top of the ICF wall (*IRC* Section R611.9).

It is important that a direct tension tie be provided from the wall to the floor or roof framing or blocking to resist loads perpendicular to the wall (this connection keeps the wall from pulling away from the floor or roof framing). This does not necessarily occur for lower Seismic Design Categories. For top bearing connections, sill plate to framing connections rely on minimum prescriptive fastening at framing perpendicular to the wall and no provision exists for a tension connection at framing parallel to the wall. For ledger-bearing connections, the direct tension ties shown in *IRC* Figures R611.8(2) through R611.8(5) are not required for lower SDCs, leaving the remaining ledger connection susceptible to cross-grain tension failure as illustrated in Figure 5-17. Very specific prescriptive anchorage requirements capable of resisting perpendicular-to-wall loads are given for sill or ledger to wood framing connections for houses in SDCs D_1 and D_2. Included are angle clips from sill to joist or blocking, continuous straps tying blocking together, and direct tension ties at ledger-bearing connections. Detailing is illustrated in *IRC* Figures R611.8(1) to 611.8(7) and R611.9. The specification of a maximum bolt size of 3/8-inch diameter for top-bearing connections is intended to favor bending of the bolt as a failure method over a brittle failure of the concrete or splitting of the wood. Larger bolt sizes should not be substituted.

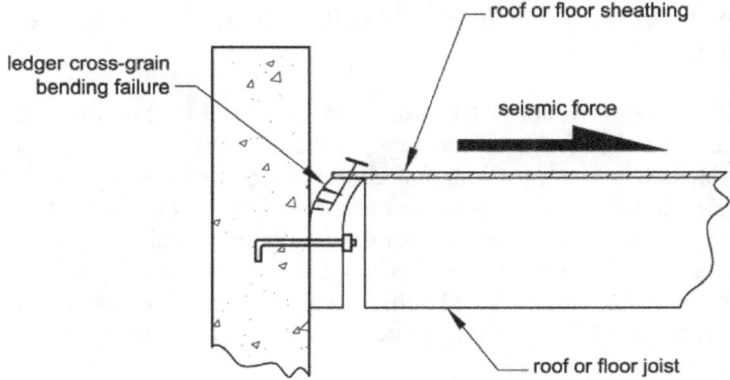

Figure 5-17 Illustration of cross-grain bending of wood ledger.

IRC Section R611.8.3 addresses requirements for floor and roof systems. The floor and roof systems act as horizontal beams carrying wind and earthquake loads to the bracing walls. The increased weight of the ICF walls results in increased earthquake load in the floor and roof systems. For SDCs A through C, use of wood structural panel sheathing is required, and the *IRC* minimum fastening schedules of 6 inches around the perimeter of each sheet and 12 inches along intermediate supports are referenced. *IRC* Section R611.8.3.1 triggers additional requirements for houses in SDCs D_1 and D_2. Minimum panel thicknesses are specified for floor and roof wood structural panel sheathing, minimum nail diameter and penetration are specified, all panel edges are required to be blocked, and the panel edge nailing is required to be 4 inches on center in SDC D_1 and 3 inches on center in SDC D_2.

IRC Section R611.9.1 has similar requirements for roof construction but also contains two notable variations. First, the requirements apply to houses in SDCs D_1 and D_2. Second, where gable end wall conditions occur, it is required that a wood structural panel sheathed attic be provided to support the gable end wall (for truss roofs, the wood structural panel sheathing can be applied to the bottom of the truss between the truss chord and the ceiling finish material) and that the wall be anchored in accordance with requirements for anchorage to floor and roof diaphragms. *IRC* Section R611.2 requires that the gable-end portion of this wall be light-framed. The writers of the ICF provisions preferred this solution over allowing increased ICF wall heights at the gable end due to the increased loads associated with concrete and the additional bracing requirements.

IRC Section R403.1 requires that exterior walls be supported on continuous concrete or masonry footings and regulates minimum footing width and depth. If interior ICF walls are provided, these too should be supported on substantial continuous footings extended to a depth that provides competent bearing.

5.5.3 Problematic Gaps Between Prescriptive and Engineered Construction

The current *IRC* requirements for ICF wall construction reveal some gaps between prescriptive and engineered construction creating an opportunity for **above-code** measures to improve performance in areas of high earthquake risk.

There are no *IRC* maximum spacing limitations between lines of ICF bracing walls corresponding to the light-frame limits of 25 and 35 feet. In light-frame construction, these spacing limitations are intended to help distribute the bracing walls in proportion to the house mass (dead load) as well as to limit the loading to the floor and roof. This helps to reduce concerns regarding house rotation due to torsional irregularities and concerns regarding irregular floor and roof system shapes. For ICF wall houses, the scope limitation to rectangular houses reduces the likelihood of rotational behavior.

> **Above-code Recommendation: Careful balancing of bracing walls around the house perimeter is recommended to further limit torsional behavior.** The maximum 60-foot house dimension will provide some limit for earthquake loads in the floor and roof. The cost of distributing the wall segments around the perimeter of the house should not result in any increased cost for construction.

For ICF wall houses, some of the irregularity limitations developed for light-frame houses have been made applicable. The concepts behind *IRC* Section 301.2.2.2.2, Irregularities 1 through 6, apply equally to an ICF wall house.

> **Above-code Recommendation:** Any of the many required or recommended measures for areas of high earthquake risk would improve performance of ICF wall houses in areas of lower earthquake risk as well as in high-wind areas. Most of the recommendations would improve house performance in high-wind events. Priorities include wall anchorage using details developed to resist out-of-plane wall loads such as those shown in *IRC* **Figures R611.8(2) to R611.8(7).**

5.5.4 Quality Control

Quality control for ICF wall construction can play a major role in the satisfactory performance of the house. Important steps in ICF construction include:

- Reinforcing should be placed properly. Unless specifically designed otherwise, reinforcing should be located as near the centerline of the ICF cavity as possible but at least within the middle third of the wall.

- Anchor bolts should be secured into place before concrete placement (not placed into concrete following placement of the concrete) to prevent air pockets from forming around the bolt which reduces its strength.

- Concrete should be consolidated as it is placed into the forms to prevent voids from forming.

Chapter 6
ROOF-CEILING SYSTEMS

Woodframe roof-ceiling systems are the focus of this chapter. Cold-formed steel framing for a roof-ceiling system also is permitted by the *IRC* but will not be discussed; rather, the reader is referred to the *AISI Standard for Cold-Formed Steel Framing – Prescriptive Method for One- and Two-Family Dwellings* (2001) for guidance. Most of the recommendations for improving the earthquake performance of woodframe roof-ceiling systems also apply to cold-formed steel construction since the systems are very similar.

6.1 GENERAL ROOF-CEILING REQUIREMENTS

Woodframe roof-ceiling systems, regardless of the pitch of the roof, form a roof diaphragm that transfers earthquake lateral loads to braced walls in the story level immediately below the roof in the same manner that floors transfer loads from interior portions of the floor to the braced wall lines of the story below. The lateral loads in the roof-ceiling are based on the mass of the roof-ceiling assembly and a portion of the mass of the walls in the story immediately below the roof.

Woodframe roof-ceilings typically consist of repetitive rafters and ceiling joists or prefabricated (engineered) trusses at a prescribed spacing. They are sheathed with either spaced solid wood boards or with wood structural panels attached to the top surface of the rafter or truss. Figure 6-1 illustrates this type of roof and ceiling framing system. Roof members also can consist of repetitive beams spaced further apart than rafters, either with or without ceiling joists.

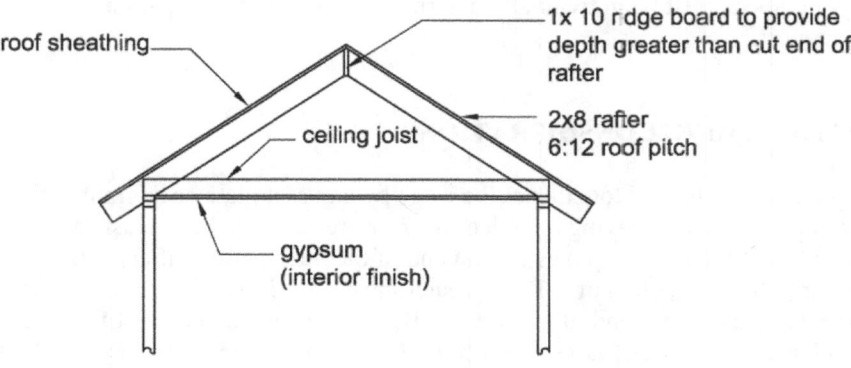

Figure 6-1 Typical light-frame roof-ceiling system.

Depending on the roof shape chosen for a house, hip and valley beam members may be needed where intersecting rafters change the direction of their span. Depending on the slope of the roof, hip and valley rafters can experience very high loads when supporting long-span rafters. Therefore, purlins are sometimes provided below and perpendicular to rafters. The purlins, in turn, are supported by a stud or post attached to a wall or beam below. Ridge boards also are often provided at the peak of a roof where sloping rafters meet. Blocking between rafters (or trusses) is used at the bearing points of rafters and ceiling joists to prevent lateral movement or rolling of the rafter. Finish materials such as gypsum board are typically applied to the bottom surface of ceiling joists or the bottom chord of a truss if the space below is to be occupied.

Rafters, purlins, ridge boards, and hip or valley members can be sawn lumber, end-jointed lumber, or any one of a variety of prefabricated (engineered) members. Examples of engineered lumber include wood I-joists or solid rectangular structural composite members such as parallel strand lumber (PSL), laminated veneer lumber (LVL), or laminated strand lumber (LSL). Roof beams and blocking can be either sawn lumber or engineered lumber.

The minimum required size and maximum span and spacing of roof rafters or beams, ceiling joists, and trusses is based on providing adequate support for dead and live vertical loads prescribed by the code. Snow loads must be considered for rafters, and attic storage must be taken into account for ceiling joists. Vertical deflection of rafters and ceiling joists is another design consideration that may limit the maximum span of these members. Rafter spans listed in prescriptive tables are based on the horizontal projection of the rafter rather than being measured along the slope, which would be a greater distance.

Tables in *IRC* Chapter 8 and similar tables in other documents such as those published by the American Forest and Paper Association (AF&PA) or engineered lumber manufacturers are available for use in selecting the proper combination of size, span, and spacing of most roof-ceiling framing members. Depending on the roof pitch, certain roof members require engineering to determine their size. For a roof pitch less than 3:12, the size of ridge boards and hip or valley members must be individually determined based on their spans and the span of the rafters they support.

6.2 SPECIAL FRAMING CONSIDERATIONS

Roof rafters must either be tied together at the ridge by a gusset plate or be framed to a ridge board. Ridge boards in roofs having a pitch of 3:12 or greater must be at least 1x nominal thickness and at least the same depth as the cut end of the intersecting rafters. Valley and hip members in a roof having a pitch of 3:12 or greater must be at least 2x nominal thickness and at least the same depth as the cut end of the rafters. Because the cut end depth of a rafter increases with increasing roof pitch, a 2x8 rafter will need a 1x10 nominal ridge board and a 2x10 nominal hip or valley member for a pitch up to 9:12. At a pitch exceeding 9:12, a 1x12 or 2x12 will be needed because the cut end of a 2x8 rafter will be greater than the 9-1/4 inch actual depth of a 1x10 or 2x10 nominal member. Figure 6-2 shows a 1x12 ridge board for a 12:12 pitch condition with the dimension for the cut end of a 2x8 rafter.

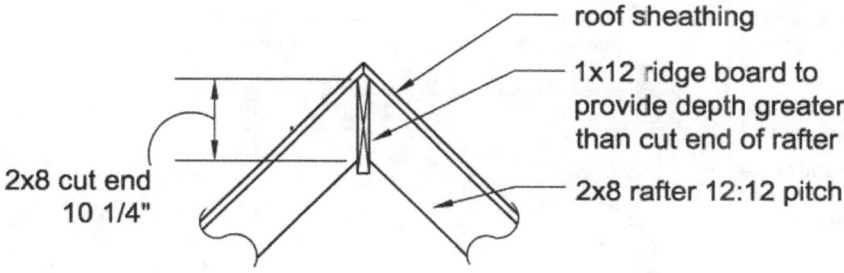

Figure 6-2 Ridge board for 12:12 pitch roof.

One additional consideration when using nominal 1x ridge boards is that ceiling joists or rafter ties are needed at the top plate of the supporting walls to prevent the walls from spreading outward. The ceiling joists or rafter ties act as a brace to resist the outward thrust of the rafters at the wall support ends of rafters. In the absence of ceiling joists or ties at the wall top plate, rafters must be supported at the ridge by a beam designed for the support of rafter loads. In addition, where a nominal 1x ridge is used and a 2x valley or hip member intersects the ridge, the valley or hip member must be supported by a stud or post attached to a bearing wall below to transfer the high loads associated with hip and valley rafters.

Above-code Recommendations: A special condition can occur at the gable ends of a roof. When the exterior wall at the gable end has its double top plate level with the low ends of the roof framing, the wall studs are not continuous to the roof sheathing along the gable edge. A section view of this condition is shown in Figure 6-3. **To provide above-code performance, the framing extending above the top plate to the level of the sloping gable end roof sheathing should be braced at regular intervals of not more than 4 feet on center at both the wall top plate level and along the top edge of the sloping roof edge.** This bracing permits the top of the exterior wall and the framing extending above to resist the lateral loads that are acting perpendicular to the wall. Without this bracing, the framing above the double top plate could be easily displaced because a hinge can form where the gable end wall framing attaches to the top plate.

Another above-code alternative to the framing shown in Figure 6-3 would be to provide wall studs that are continuous to a sloping double top plate located just below the roof sheathing (balloon framing). When continuous studs are provided, there is no weak location for a hinge to form. However, when this method is used, the required stud size, spacing, and maximum height must comply with *IRC* Table R602.3.1 or the wall studs will need to be engineered by a design professional. Bracing a gable end wall also is important for providing resistance to high winds, especially for roofs with a steep pitch.

FEMA 232, Homebuilders' Guide

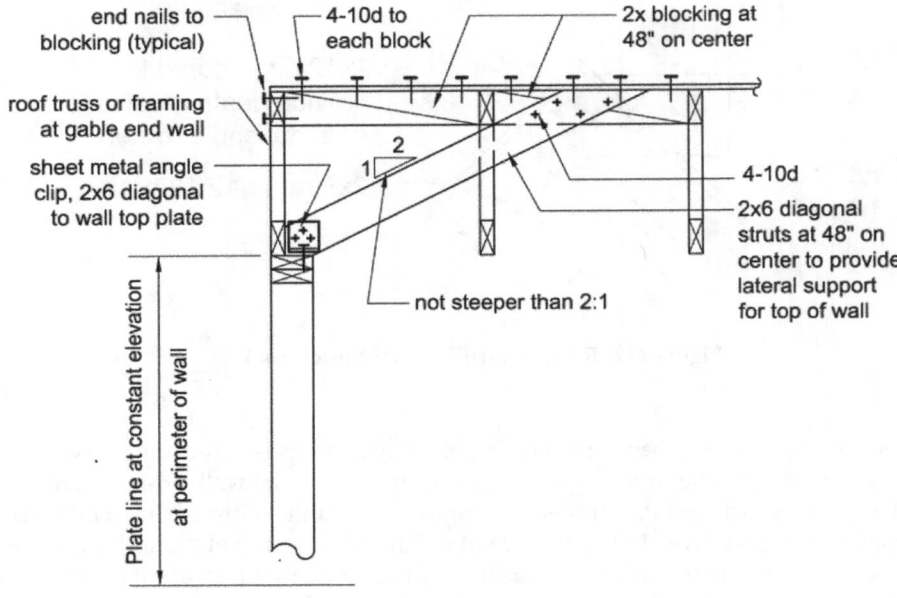

Figure 6-3 Gable end wall or gable truss bracing.

6.3 BLOCKING AND LATERAL LOAD PATHS FOR ROOF SYSTEMS

Rafters and ceiling joists having a nominal depth-to-thickness ratio exceeding 5:1 (e.g., 2x10) need blocking at their points of bearing to prevent them from rotating or displacing laterally from their intended position. Rotation loads on rafters occur when the roof sheathing is resisting lateral loads perpendicular to the rafter because these loads are actually trying to move the top edge of the rafter sideways. Preventing rotation is typically accomplished by installing full-depth solid blocking along wall top plates between rafters and ceiling joists. Figure 6-4 illustrates blocking installed between adjacent pairs of a ceiling joist and rafter that are bearing on an exterior wall.

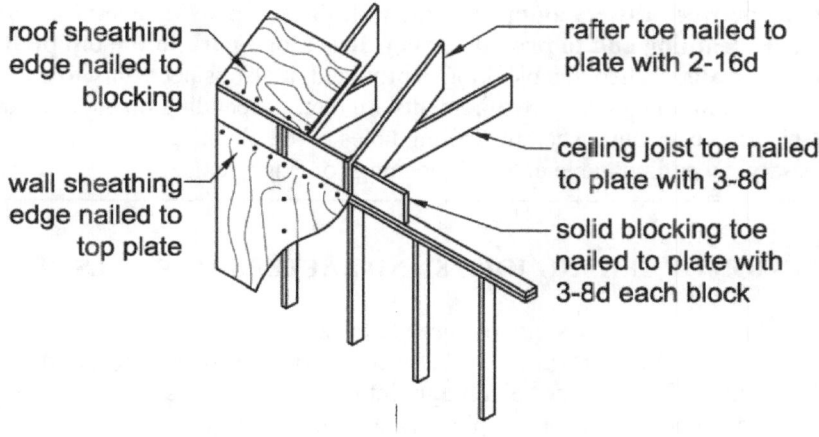

Figure 6-4 Blocking at rafters to exterior wall.

However, when the nominal size of a ceiling joist or rafter is 2x10 or smaller, blocking over the exterior wall may be omitted. This would be likely to occur where the roof overhang is enclosed by a soffit. Without an enclosed soffit, blocking is typically provided between rafters along the exterior wall line regardless of the rafter size to serve as a barrier between the outside and the attic space.

Because attics (or spaces between rafters where ceiling finish is attached to the bottom of the rafter) require ventilation openings, the spaces between rafters along exterior walls are typically used to provide some portion (if not all) of the required attic ventilation opening. When blocking is installed, ventilation openings often are provided by drilling holes in the blocking; when blocking is not required, ventilation can be provided through the entire opening space between rafters.

Although blocking may not be required for 2x10 and smaller rafters, there still must be a load path for lateral loads in the roof sheathing to reach the exterior braced walls immediately below the roof. The most direct load path is for the roof sheathing to be edge nailed to blocking between each rafter. That blocking is then nailed to the wall top plate with three 8d nails per rafter space as prescribed in *IRC* Table 602.3(1). This load path is shown in Figure 6-4. Although alternate load paths are certainly possible, the one shown is the most direct and is essentially the same load path the *IRC* provides between a floor and the braced walls below that floor.

> **Above-code Recommendation:** In Seismic Design Categories D_1 and D_2, blocking is recommended between rafters along exterior wall lines to provide a surface for the edge nailing of roof sheathing and to provide a very direct load path to the top plate of the exterior braced walls. When this blocking is provided, it also is necessary to ensure that minimum attic ventilation opening requirements are met. Depending on the attic area being ventilated, this can be accomplished by drilling holes in the blocking and, when more opening area is necessary, by adding gable end wall openings or ridge vents.

6.4 CONNECTION OF CEILING JOISTS AND RAFTERS TO WALLS BELOW

Ceiling joists and rafters (or trusses) are required to be connected to the top plate of supporting walls as specified in *IRC* Table R602.3(1). These connections also provide a portion of the load path that transfers loads from the roof diaphragm into the braced walls below. Ceiling joists require a toe-nailed connection to the top plate using three 8d box or common nails. Rafters also require a toe-nailed connection to the top plate using two 16d box or common nails. Blocking installed between the rafters or ceiling joists requires a toe-nailed connection to the top plate using a minimum of three 8d box or common nails in each block. Toe nailing must be done correctly if the needed transfer of loads is to occur; therefore, ensure that the nails do not split the wood. In high-wind areas, light-gage steel connectors often are used in place of these toe-nailed connections. The use of commercially available light gage steel connectors in place of toe nails can reduce wood splitting and provide more reliable load transfer. Additional information on proper toe-nailing installation is illustrated in Figure 4-8.

6.5 ROOF SHEATHING

Wood boards installed either perpendicular or at an angle to the rafters (or trusses) or wood structural panels can be used as roof sheathing. *IRC* Table R803.1 specifies the minimum thickness for wood board roof sheathing for various spacings between roof rafters, trusses, or beams.

> **Above-code Recommendation:** Solid wood board roof sheathing is rarely used in modern housing construction except perhaps along roof eave overhangs. When wood boards are used and are installed perpendicular to rafters they provide a very weak diaphragm with little stiffness. **As a result, solid wood board sheathing installed perpendicular to rafters should not be used in Seismic Design Categories C, D_1, and D_2.** Wood boards installed diagonally would provide much better diaphragm capacity but are very rarely used in modern housing construction.

Wood structural panel roof sheathing is the most common roof sheathing used in current construction. The minimum thickness is based on rafter (or truss) spacing and the grade of sheathing panels selected. *IRC* Table R503.2.1.1(1) is used to determine the minimum required thickness for wood structural panel roof sheathing materials for a variety of rafter spacings. For roofs, the short direction panel joints between wood structural panels can be either staggered or not staggered. Typical wood structural panel roof sheathing installation using staggered joints is illustrated in Figure 6-5.

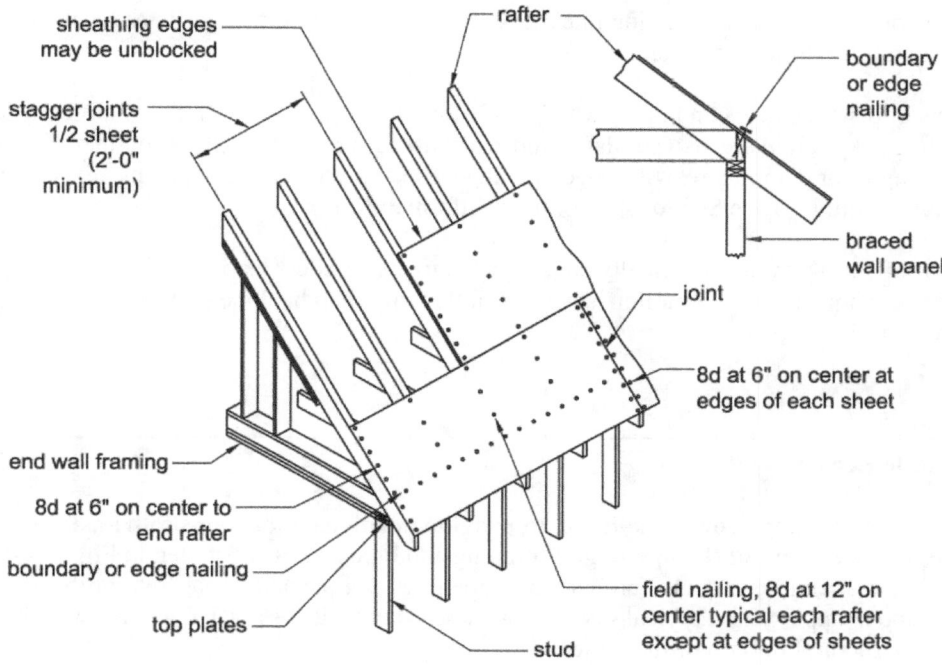

Figure 6-5 Typical roof diaphragm sheathing nailing when wood structural panels are used.

When the roof covering consists of wood shingles or shakes, *IRC* Chapter 9 permits the use of spaced wood boards as roof sheathing. However, spaced wood board sheathing is not permitted in Seismic Design Category D_2. This limitation in SDC D_2 is intended to provide a stiffer and stronger roof diaphragm that will resist the larger lateral loads anticipated in SDC D_2. When wood boards are used, they typically are nailed to each rafter with only two 8d box or common nails. Wood structural panel sheathing results in the stiffest roof diaphragms, and the panels typically are fastened to the rafters with 8d box or common nails spaced at 6 inches along supported edges and 12 inches in the field of the panel. When spaced wood boards are used, even fewer nails are provided than with solid wood board sheathing and even less lateral resistance results.

> **Above-code Recommendation:** Spaced wood board sheathing is not recommended for use in Seismic Design Categories C and D_1.

6.6 LATERAL CAPACITY ISSUES FOR WOOD FRAMED ROOFS

The lateral capacity of a roof diaphragm sheathed with wood structural panels is based upon the same five factors as floor diaphragms. (See the discussion in Section 4.7 of this guide for more information on the effects of sheathing thickness, fastener size and spacing, use of blocking, and layout of wood structural panel sheathing.)

The effects of distance between braced wall lines below the roof and the effects of large roof openings (e.g., skylights) are also similar to those discussed for floors. The recommendations for reinforcing floor diaphragms with large openings also apply when roof openings exceed the code maximum limits. (See Section 4-7 for more information.)

Although roof-ceiling systems typically are subjected to the smallest loads when a house experiences earthquake ground motion, several simple things can be done to improve the performance of the system.

Above-code Recommendations:

Reinforce the framing around skylights to provide positive connections to transfer the diaphragm loads around the opening. Strapping and blocking as illustrated in Figure 4-11 can be used to strengthen the roof around large openings and the additional cost for the blocking and strapping for a typical skylight opening would add less than 0.4 percent to the cost of the structural portion of the project.

Make sure all rafters and ceiling joists are blocked at all locations where they are in contact with a top plate of a wall below (alternately a rim joist may be used). For the exterior walls of the model house, this would cost 0.8 percent of the structural portion of the project. If interior walls were also blocked, an additional cost of 0.7 percent of the structural portion of the project would be incurred.

Glue the sheathing to the roof framing in the same way most floor sheathing is installed and then block the roof as well. Construction adhesives are typically used to prevent squeaky floors but they also strengthen and stiffen. The cost of adding adhesive to the roof system for the model house in this guide would be 1.8 percent of the structural system, and adding blocking to the roof system would add another 3.4 percent to the cost of the structural system.

6.7 QUALITY CONTROL

The most important item to monitor for quality of the roof diaphragm system is to ensure that the nails used to attach the sheathing to the framing are driven flush with the top of the sheathing and not overdriven and counter sunk into the sheathing materials. Overdriving sheathing nails has been shown to reduce the strength of shear walls and diaphragms. Evidence of a 40 to 60 percent loss in strength of the shear wall and diaphragm has been observed in laboratory tests of assemblies with overdriven nails. (See Section 5.1.4 of this Guide for more discussion on the effects of overdriven nails and how pneumatic tools can be altered to correct for this error.)

To ensure the quality of the roof-ceiling system, make sure blocking is installed correctly so that the wood framing is not split by the toe nailing.

If adhesives are used to attach the sheathing, attention needs to be paid to the time between the application of the adhesive and when the nails are driven to hold the sheathing in place. Especially in hot weather, the adhesive tends to skin over or cure on the surface quickly, which reduces the adhesion between the glue and the sheathing. Check the time allowed for each specific product used.

Chapter 7
CHIMNEYS, FIREPLACES, BALCONIES, AND DECKS

This chapter provides an overview of the *IRC* provisions for earthquake-resistant design and construction of chimneys, fireplaces, balconies, and decks in houses. **Above-code** recommendations for improved earthquake performance are provided.

7.1 CHIMNEYS AND FIREPLACES

IRC Chapter 10 presents requirements for masonry fireplaces and chimneys and for factory-built fireplaces and chimneys enclosed in framing. The provisions of the *IRC* are intended for the moderately sized fireplaces and chimneys commonly found in houses.

> **Above-code Recommendation:** Where fireplaces or chimneys are large or oddly configured, an engineered design is encouraged in order to fully address the design of the chimney and fireplace and their influence on the house.

7.1.1 Masonry Chimneys and Fireplaces

Although the *IRC* permits construction of masonry fireplaces and chimneys in earthquake-prone regions, masonry chimneys are particularly vulnerable to earthquake damage, and such damage has occurred in most moderate to severe U.S. earthquakes (Figures 7-1 and 7-2). Masonry fireplaces and chimneys can be heavy and rigid, and many chimneys in existing houses also are brittle. The movement of the fireplace and chimney in response to earthquake ground motions can be significantly different from the movement of the light-frame house itself, creating the potential for damage to both the chimney and the house.

Figure 7-1 Chimney damage.
Photo Courtesy National
Information Service for
Earthquake Engineering,
University of California, Berkeley

Figure 7-2 Chimney damage in Northridge earthquake.

The *IRC* triggers requirements for masonry fireplace and chimney reinforcing steel and anchorage to floors, roofs, and ceilings for houses in SDCs D_1 and D_2. Although these requirements cannot completely eliminate the possibility of damage to the fireplace and chimney in an earthquake, their use permits a chimney to better withstand earthquake loads and should lessen the falling hazard posed by a damaged chimney. Although it may be possible with systematic engineering design to mitigate the damage often seen in masonry chimneys and fireplaces, the lower weight and greater flexibility of factory-built fireplaces and chimneys make them the better choice for light-frame houses in earthquake-prone areas.

A substantial footing is necessary if a fireplace and chimney is to perform well under any type of loading. The footing should extend to a depth not less than that of surrounding footings. *IRC* Section R1003 contains minimum footing requirements.

IRC Section R1003.3 provides requirements for masonry chimney and fireplace reinforcing steel. The minimum amount of vertical reinforcing steel is four No. 4 bars for a chimney up to 40 inches wide (a depth of approximately 24 inches is common). An additional two No. 4 vertical bars are required for each additional flue or each additional 40 inches of width. Where the reinforcing bars cannot run full height, a lap splice of not less than 24 inches is needed. Grout, continuous from the footing to the top of the chimney, must surround the reinforcing steel. For horizontal reinforcing, a minimum of 1/4-inch ties at not more than 18 inches on center is required in the mortar joints. *IRC* Section R1003.3 also cites the Section R609 requirements for grouted masonry discussed in Chapter 5 of this guide. Proper grouting and consolidation around the reinforcing steel is needed in order for the reinforcing and anchorage to contribute to earthquake resistance. Lack of grout and poorly consolidated grout are common contributors to earthquake damage. It is also important to note that *IRC* Section R609 prohibits the use of Type N masonry mortar in SDCs D_1 and D_2.

Anchorage of the masonry chimney to the framing at each above grade floor, roof, and ceiling level is necessary. *IRC* Section R1003.4 provides anchorage requirements applicable in SDCs D_1 and D_2. Steel straps not less than 3/16-inch by 1-inch are required to extend a minimum of 12 inches into the chimney masonry, hook around outer reinforcing bars, and extend not less than 6 inches beyond the hook. Chimney anchorage locations are illustrated in Figures 7-3 and 7-4. The *IRC* provisions specify anchorage to a minimum of four ceiling or roof joists with not less than two 1/2-inch bolts. This description does not give details of the intended configuration and does not address framing parallel to the chimney wall. Figures 7-5 and 7-6 illustrate implementation of this anchorage provision with detailing consistent with industry recommendations (MIA, 1995) and earlier *Uniform Building Code* provisions. Anchorage in general conformance with Figures 7-5 and 7-6 should be consistent with the intent of the *IRC*.

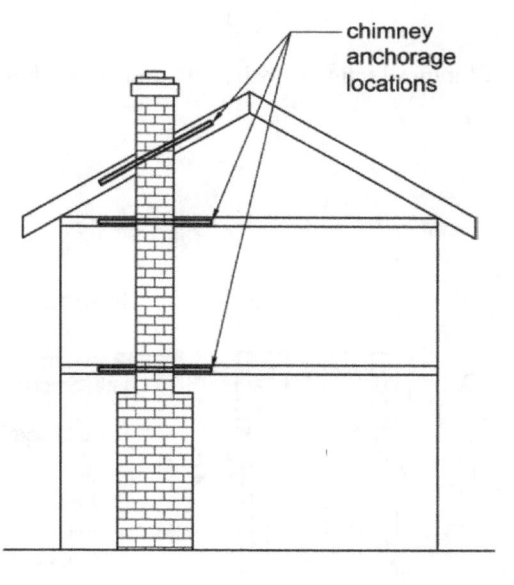

Figure 7-3 Locations for earthquake anchorage of masonry chimney at exterior house wall.

FEMA 232, Homebuilders' Guide

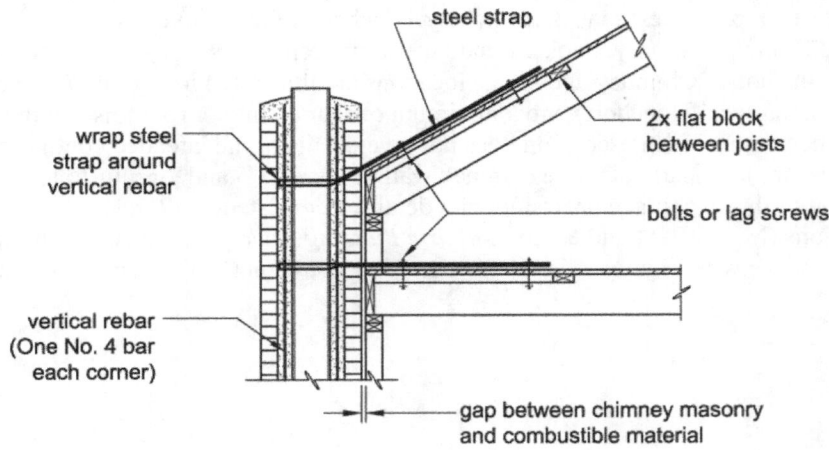

Figure 7-4 Chimney section showing earthquake anchorage.

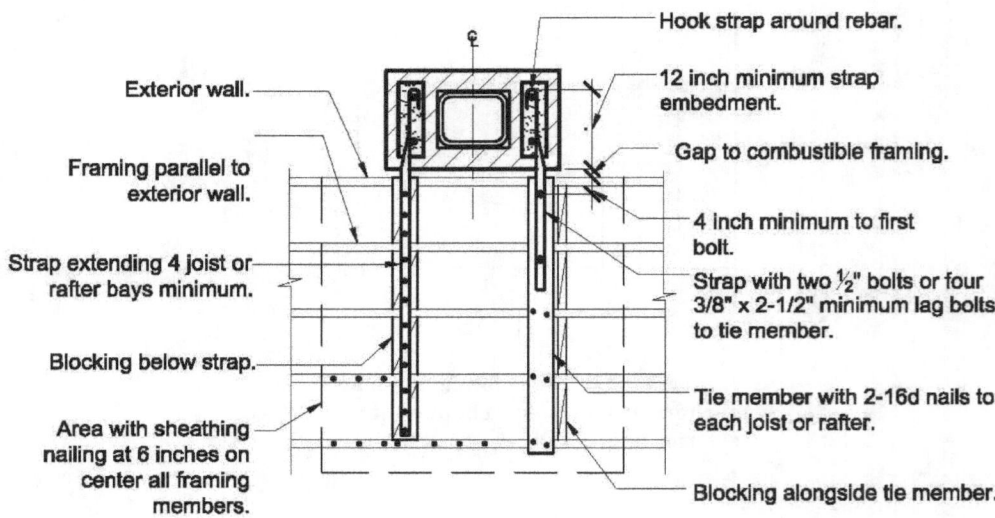

Figure 7-5 Anchorage detail for framing parallel to exterior wall.

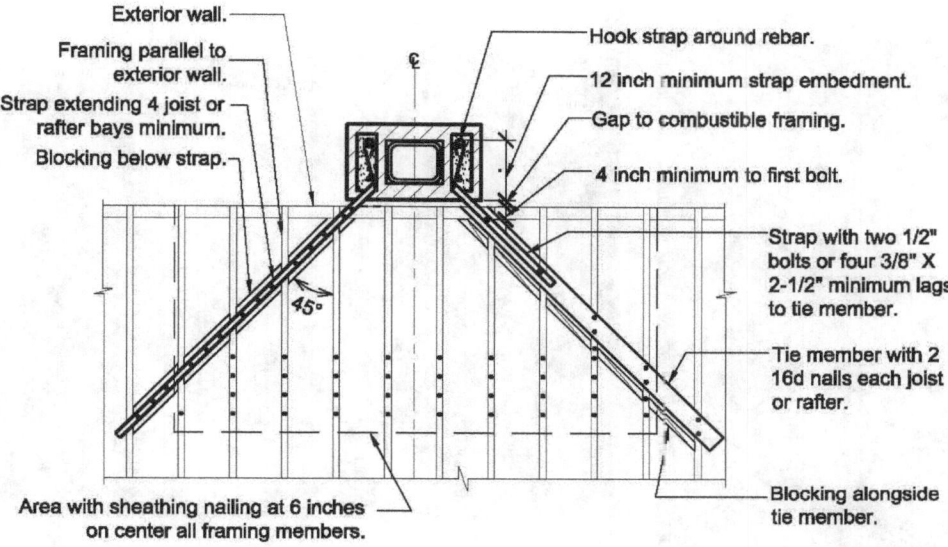

Figure 7-6 Anchorage detail for framing perpendicular to exterior wall.

To reduce the fire hazard with respect to the surrounding wood structure, *IRC* Section R1001.15 requires a clearance of 1 to 2 inches (depending on configuration) between combustible framing materials and the masonry chimney. Detailing of earthquake-resistant anchorage at the floor, ceiling, and roof levels needs to maintain this required clearance with only the steel straps extending across this clearance gap.

Even when fireplaces and chimneys show no signs of damage after an earthquake, the masonry or flue liner may have cracked, and inspection before reuse is recommended.

7.1.2 Factory-Built Fireplaces and Flues

Factory-built fireplaces and flues generally are installed within light-frame fireplace enclosures and chimneys. During an earthquake, the deflections of the light-frame enclosure are compatible with those of the house, which greatly reduces the potential for damage. Detailing of framing anchorage, however, is important. At exterior walls, as shown in Figure 7-7, the framing for the chimney disrupts the typical wall and roof framing. Wall top plates often are discontinued and studs are balloon framed to the top of the light-frame chimney. Measures should be taken to restore the continuity of the top plates and to anchor the fireplace/chimney wall framing to the floor and roof. Figure 7-8 shows how one light-frame chimney without sufficient connections behaved. No specific requirements for this construction currently exist in the *IRC*. Clearances to combustible wood framing remain important with factory-built fireplaces and chimneys and are typically addressed in the installation instructions.

Figure 7-7 Factory-built flue and light-frame enclosure.

Figure 7-8 Collapsed factory-built chimney, light-frame enclosure, and deck after the San Simeon earthquake. Photo courtesy of Josh Marrow, Simpson Gumpertz and Heger Inc.

> **Above-code Recommendations:** Factory-built fireplaces and flues typically are installed within light-frame fireplace enclosures and chimneys that generally perform well during earthquakes; therefore, their use in the higher Seismic Design Categories is recommended. However, special attention should be given to the detailing of the framing anchorage and to compliance with clearances to combustible wood framing addressed in installation instructions.

Use of stone or masonry veneer increases the weight of the light-frame chimney, thereby increasing earthquake loads proportionately. Particular care should be taken to tie the wood framing into the floor and roof when veneer is used. Veneer attachment to the framing should be in accordance with *IRC* Chapter 7. See Section 5.2 of this guide for a discussion of veneer attachment.

> **Above-code Recommendation:** Use of reinforcing steel and chimney anchorage are recommended to improve the performance of fireplaces and chimneys across all Seismic Design Categories and particularly in SDC C. Adding reinforcement and anchorage for the chimney on the model house used in this guide would increase the cost of the structural portion of the house by approximately 2 percent, which is approximately 0.5 percent of the total cost.

> **Above-code Recommendation:** Both masonry and factory-built fireplaces and chimneys result in increased weight and earthquake loading as well as discontinuities in house configuration. **Additional bracing walls in both directions in the vicinity of the fireplace are recommended to resist the additional earthquake load.** This will result in a reduction in the amount of window opening available, but the cost of adding the wall will likely be offset by the reduced cost for windows.

7.2 BALCONIES AND DECKS

Balconies and decks often are prominent features of modern residential construction and, for many people, add much desired living space (see Figure 7-9). Although the *IRC* contains some provisions addressing balconies and decks, the earthquake resistance of balconies and decks has not been systematically considered in the development of the *IRC* provisions. This section addresses three aspects of balconies and decks that need consideration:

- The effect of added floor area beyond braced wall lines,
- Anchorage for earthquake loads, and
- Vertical support.

7.2.1 Added Floor Area Beyond Braced Wall Lines

The bracing provisions of the *IRC* reflect the need to support the floor area within the exterior braced wall lines. The addition of balconies and decks creates additional weight and increases earthquake loads, a fact that was not envisioned when required bracing lengths were determined. Although the addition of a small balcony or deck is not likely to greatly affect the earthquake performance of a house, the addition of a large balcony or deck may. Further, balconies and decks tend to concentrate the added earthquake load on one side of the house and one braced wall line. This can contribute to rotational behavior (see the discussion of plan irregularities in Section 2.3 of this guide) and concentrations of damage.

Two *IRC* sections limit balcony and deck size:

- *IRC* Section R502.3.3 addresses floor framing cantilevers, including exterior balconies. *IRC* Table R502.3.3(2) specifies permitted cantilevers as a function of framing size and spacing and live load. Permissible cantilevers range up to a maximum of 6 feet (see additional discussion of back-span connections in Section 4.3 of this guide).

- For SDCs D_1 and D_2, *IRC* Section R301.2.2.2.2, Item 2, requires braced wall lines on all edges of a floor or roof. The exception to Item 2 permits floors that do not support braced wall panels to extend up to 6 feet beyond a braced wall line.

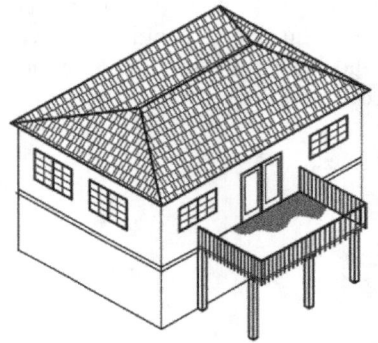

Figure 7-9 Decks and balconies in residential construction.

Balconies extending more than 6 feet beyond the exterior house wall in all Seismic Design Categories fall outside of the framing provisions of the *IRC* and require engineered design.

Where balconies or decks extend more that 6 feet beyond the house exterior, additional lines of bracing for earthquake and wind loads are sometimes provided. When this approach is taken, it is important that loading in both the longitudinal and transverse directions be considered. In addition, when both the house and supplemental bracing are used to support the balcony or deck, the load-deformation behavior of the bracing system must be compatible with that of the house. Where this is not the case, it is best to completely separate the balcony or deck from the house and provide a gap large enough to permit independent movement.

7.2.2 Anchorage for Earthquake Loads

Where a balcony or deck is laterally supported by the house, adequate connection to the house is key to good earthquake performance. The *IRC* includes two provisions that address anchorage to the house:

- *IRC* Section R311.2.1 requires that the connections used to attach exterior balconies, stairs, and similar exit facilities to the rest of the structure provide for resistance to both vertical and lateral loads, but the magnitude of loads to be resisted is not specified. Use of toe nails or nails subject to withdrawal is prohibited.

- *IRC* Section R502.2.1 provides similar requirements for decks.

Because balcony framing generally has a back span that extends into the interior of the house, adequate connection generally is not an issue; however, care should be taken to ensure that the back span is adequately fastened to the floor sheathing or to lapped floor framing and that blocking is provided where the cantilever bears on the exterior wall.

> **Above-code Recommendation:** Deck framing, on the other hand, does not generally have inherent continuity into the house. Attachment only to the floor band (rim) joist is not sufficient and will result in failures between the band joist and the rest of the floor system. **Engineered design of connections is recommended as is use of a positive connection, such as the hold-down device shown in Figure 7-10.**

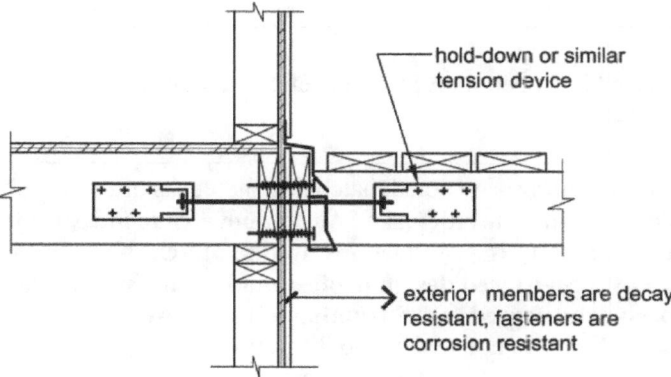

Figure 7-10 Hold-down device providing positive connection of deck framing to house framing.

7.2.3 Vertical Support Issues

Although beyond the scope of earthquake resistance, two issues – vertical load connections and moisture and decay – related to vertical support of decks and balconies are of enough significance to warrant discussion.

Deck construction is notoriously problematic and typically at least one deck collapse occurs somewhere in the United States every week. Inadequate connection between the deck and the house for vertical loads is the biggest problem. Where an engineered design is not provided, connections for vertical loads very often are inadequate. Inadequate connection for lateral loading also can be a contributor. Attention to adequacy of vertical load connections is imperative for safety.

> **Above code Recommendation: Providing a line of vertical support (posts and beams) alongside the exterior house wall can help to reduce the load on the deck-to-house connection.**

The interface between a balcony or deck and the exterior house wall is critical for the waterproofing system. Penetration of moisture at this interface can endanger not only the capacity of the connection but also the interior and exterior framing members. *IRC* Section R319.1.2 requires that the joints of exterior balconies be designed such that moisture will not collect in the connection area or that the connection be otherwise protected from moisture.

> **Above-code Recommendation: The joints of exterior decks should be designed such that moisture will not collect in the connection area or that the connection is otherwise protected from moisture.**

The resource list provided in Appendix E includes references addressing deck connections and moisture and decay issues.

> **Above-code Recommendation:** As noted, balconies and decks often are subjected to significant lateral loads during an earthquake. **As an above-code measure for houses located in SDCs C, D_1 and D_2, the connections used to attach the deck to the house should be designed by a registered design professional to ensure that the loads acting on the deck are properly transferred to the framing of the house.** In addition, the lateral bracing of the deck for the sides not attached to the house and for all sides of free-standing decks must be sufficient to prevent a torsional response.

Chapter 8
ANCHORAGE OF HOME CONTENTS

8.1 GENERAL

Anchorage of home contents can greatly reduce the risk of injury, property loss, and interruption of home use as a result of an earthquake. Anchorage is particularly recommended for large and heavy items such as water heaters, bookcases, and file cabinets and for items that could cause injury if they fell (e.g., items on shelves above a bed). Other priority items to anchor or otherwise restrain include wood stoves, similar heating appliances, and outside fuel tanks, all of which pose a fire risk.

Anchorage of other home contents will further reduce disruption following an earthquake. Measures to anchor computers and televisions range from very simple home fixes to specialized restraint systems available from a number of manufacturers. Locks on kitchen and china cabinets can help reduce the spilling and breakage of contents during an earthquake. Measures are available to secure a wide range of other home contents including pictures, mirrors, fragile objects, and fire extinguishers.

Section 8.2 provides guidance on the anchorage of water heaters. Section 8.3 addresses anchorage of a number of items using excerpts from FEMA 74, *Reducing the Risks of Nonstructural Earthquake Damage* (FEMA, 1994).

8.2 WATER HEATER ANCHORAGE

When not properly anchored, water heaters can fall over, resulting in a fire hazard and water damage. Bracing is required for new water heater installations and is recommended as a top priority for existing installations. *IRC* Chapters 20, 24, and 28 contain some provisions for the installation of water heaters but do not specifically address anchorage. Kits for bracing water heaters are available at many hardware stores. As an alternative to kits, Figures 8-1 and 8-2 provide example details for anchorage using plumber's tape (24-gage by 3/4-inch minimum steel straps) and electrical conduit. These figures are adapted from *Guidelines for Earthquake Bracing of Residential Water Heaters* (California Division of the State Architect, 2002); consult this publication for detailed installation instructions and additional cautions and limitations.

Note that Figures 8-1 and 8-2 are applicable only to water heaters with a maximum capacity of 52 gallons with two strap locations and with a maximum capacity of 75 gallons with three. Some jurisdictions place additional limits on water heater bracing details. Where water heaters are installed on a platform, the water heater base should be attached to the platform and the platform should be anchored to the floor. Placing water heaters in metal pans to retain any spilled water is a possible precaution in addition to bracing. The *IRC* requires that water heaters be placed in pans where loss of water would cause damage. Required clearances to walls and combustible construction need to be maintained; the water heater UL listing and local jurisdiction requirements should be verified prior to moving or installing a water heater.

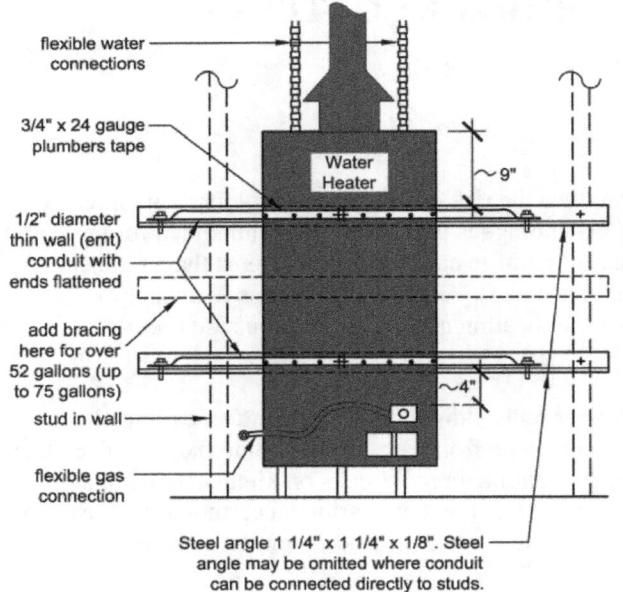

WATER HEATER WALL BRACING

This detail illustrates earthquake bracing for typical residential water heater systems (30 to 75 gallon capacity) braced to a flat wall. See wall bracing elevation at left. See second page for wall bracing plan and details.

Where the water heater is installed on a raised platform, attach water heater to platform, and platform to floor.

Do not move water heater closer to wall or combustible materials without verifying that minimum required clearances to wall and combustible materials are still met.

See text and referenced documents for additional cautions and limitations, and other available details.

Plumbers tape is also called galvanized steel hanger strap, and must be not thinner than 24 gauge.

The illustrated bracing detail is based on California Division of the State Architect *Guidelines for Earthquake Bracing of Residential Water Heaters*.

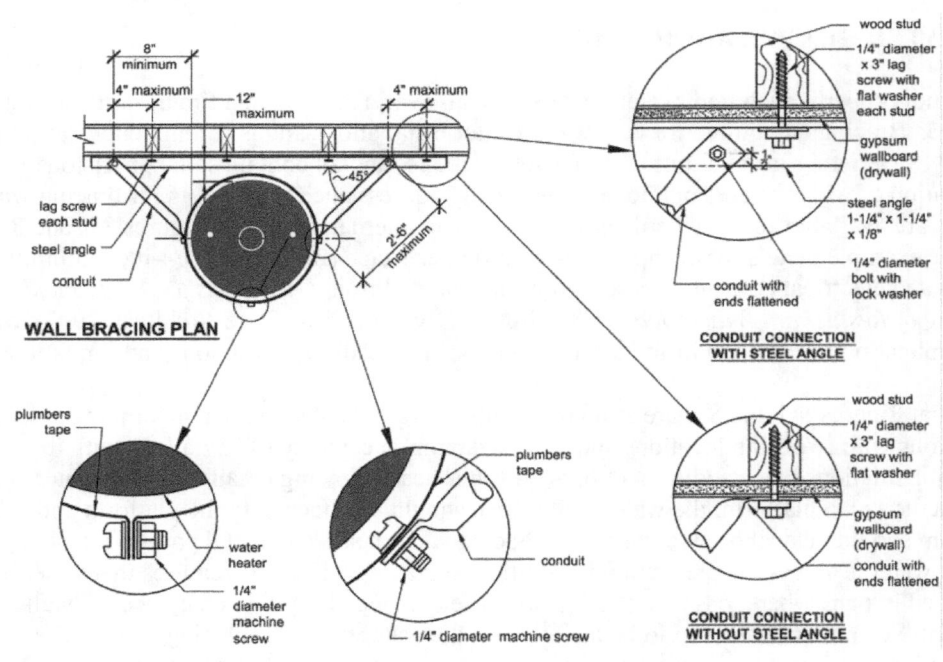

Figure 8-1a and 8-1b Securing a water heater with wall bracing.

Chapter 8, Anchorage of Home Contents

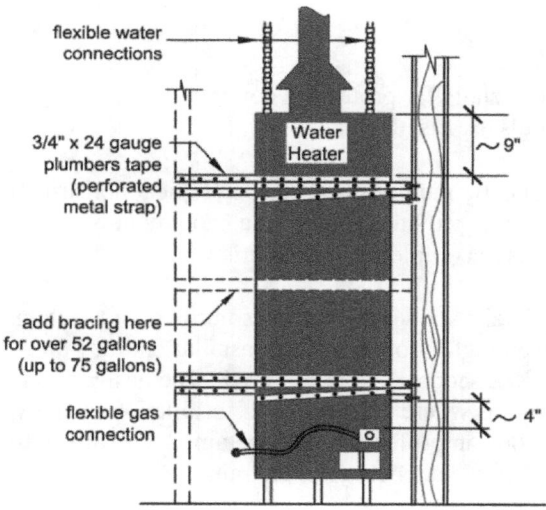

WATER HEATER CORNER BRACING

This detail illustrates earthquake bracing for typical residential water heater systems (30 to 75 gallon capacity) installed at a corner. See corner bracing elevation at left. See second page for corner bracing plan and details.

Where the water heater is installed on a raised platform, attach water heater to platform, and platform to floor.

Do not move water heater closer to wall or combustible materials without verifying that minimum required clearances to wall and combustible materials are still met.

See text and referenced documents for additional cautions and limitations, and other available details.

Plumbers tape is also called galvanized steel hanger strap, and must be not thinner than 24 gauge.

The illustrated bracing detail is based on California Division of the State Architect *Guidelines for Earthquake Bracing of Residential Water Heaters*.

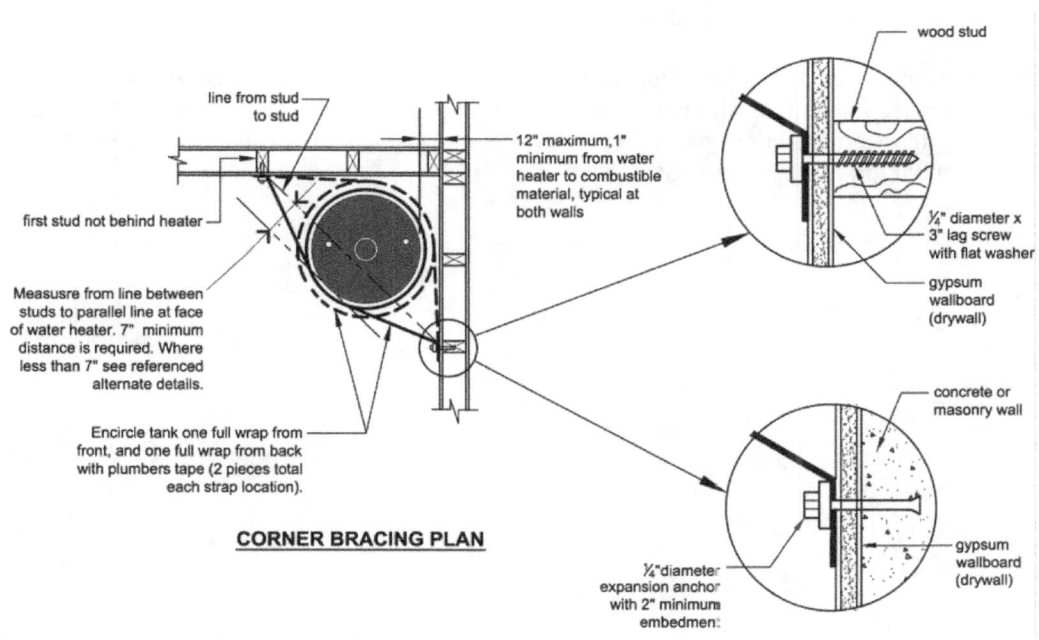

Figure 8-2a and 8-2b Securing a water heater with corner bracing.

8.3 SECURING OTHER ITEMS

The examples included here show representative details for protecting common items from earthquake damage. Two different types of details are discussed:

- Do-it-yourself methods, which are simple generic methods for securing typical nonstructural items found in the home. Enough information is provided to permit a handyman with common tools and readily available materials to complete an installation.

- Engineered methods, which are schematic details showing common solutions for the items in question. These sketches do not contain enough information for installation; they are provided here primarily as an illustration of the scope of work required. The designation "Engineering Required" has been used for items where do-it-yourself installation is likely to be ineffective. FEMA 74 recommends that design professionals be retained to evaluate the vulnerability of these items and design appropriate anchorage or restraint solutions, particularly where safety is an issue.

Cost estimates are provided with the details in this section as a rough guide for planning or budgeting purposes. The values are intended to cover the cost of materials and labor. They do not include allowance for architectural or engineering fees, permits, special inspection, etc. These estimates represent a professional opinion based on information available at the time FEMA 74 was published in 1994 so current actual construction costs may vary significantly, depending on the timing of construction, changes in conditions, the availability of materials, regional cost variations, and other factors.

FEMA 74 provides several pages of installation notes that will be of assistance to users interested in selection and installation of fasteners appropriate to construction materials of the house (www.fema.gov/hazards/earthquakes/nehrp/fema-74.shtm). Reproductions of FEMA 74 pages are shown in Figures 8-3a through 8-3f.

Chapter 8, Anchorage of Home Contents

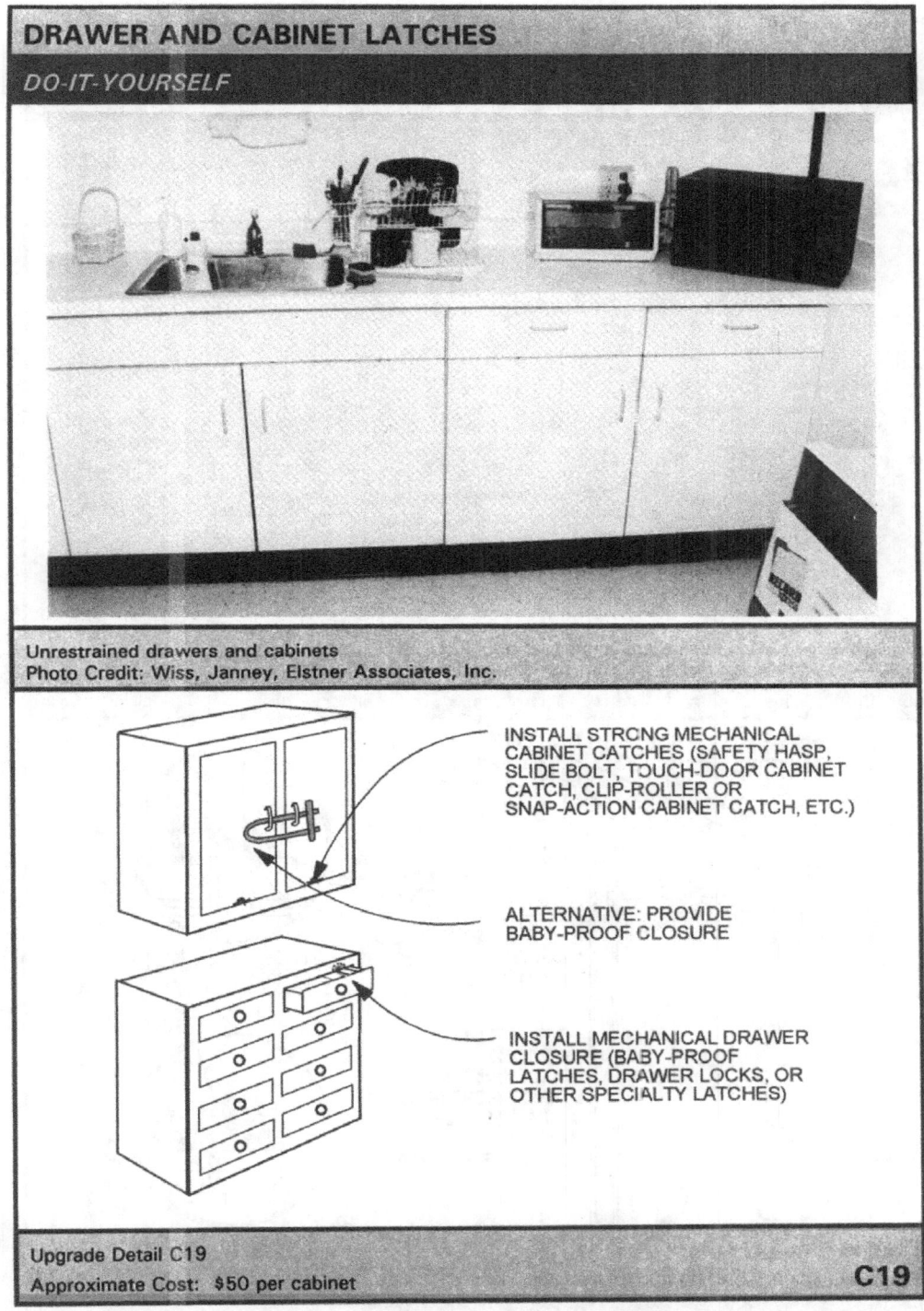

Figure 8-3a Reproduction of FEMA 74 suggestions for securing drawer and cabinet latches.

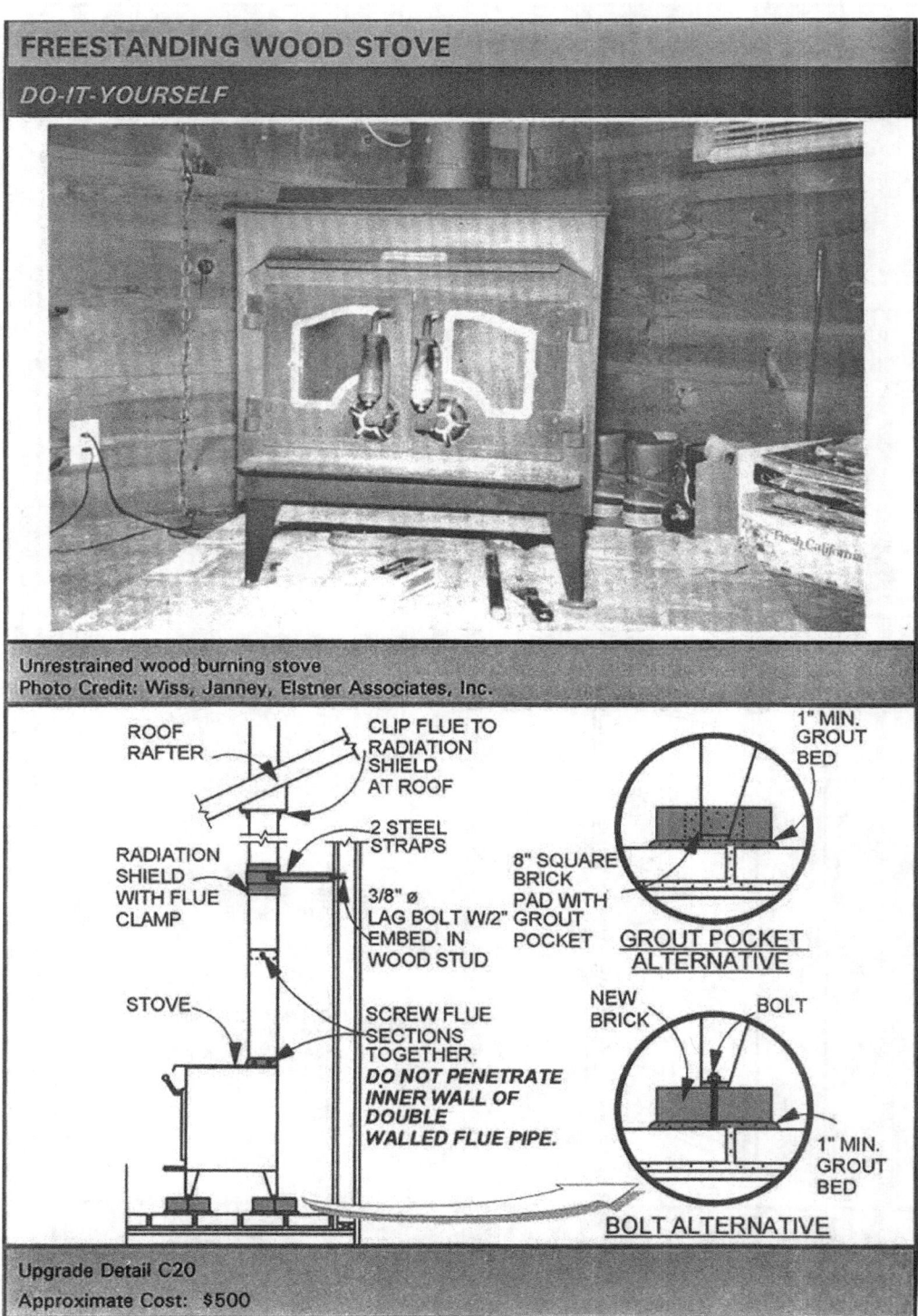

Figure 8-3b Reproduction of FEMA 74 suggestions for securing a freestanding wood stove.

Chapter 8, Anchorage of Home Contents

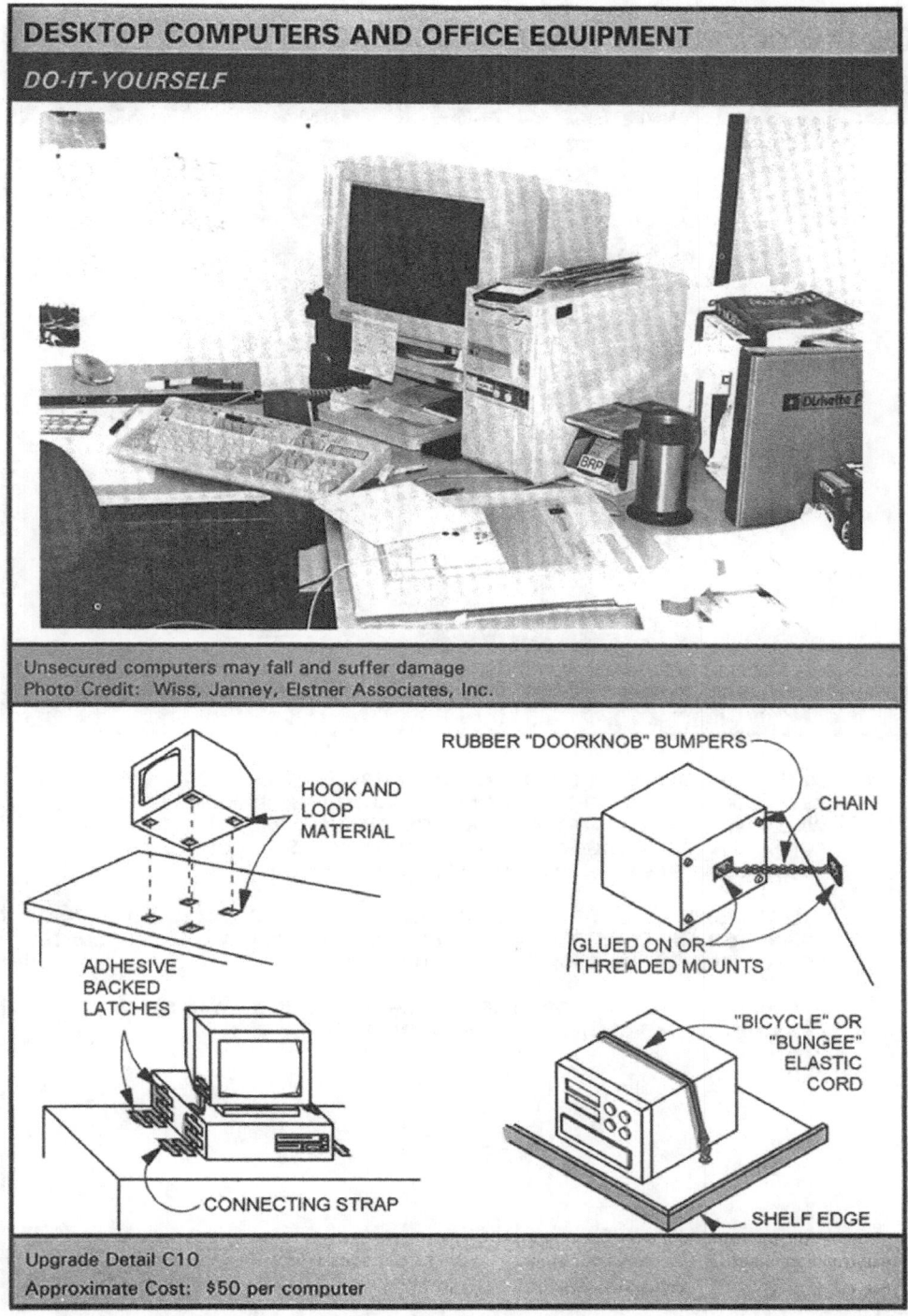

Figure 8-3c Reproduction of FEMA 74 suggestions for securing desktop computers and office equipment.

FEMA 232, Homebuilders' Guide

FREESTANDING WALLS OR FENCES
ENGINEERING REQUIRED

Earthquake Damage: 1994, Northridge, California
Photo Credit: Robert Reitherman

- MANY MILES OF POORLY CONSTRUCTED CONCRETE MASONRY UNIT (CMU) FENCES WERE DAMAGED DURING THE 1994 NORTHRIDGE EARTHQUAKE. COLLAPSE OF WALLS WITH INADEQUATE OR ABSENT REINFORCING AND/OR FOUNDATIONS WAS COMMON IN NORTHRIDGE AND SYLMAR. IN MANY CASES, MOST OF THE SIDEWALK WAS COVERED WITH DEBRIS, AS SHOWN ABOVE.

- FREESTANDING WALLS OR FENCES BUILT OF CMU, BRICK, OR STONE NEED TO BE ENGINEERED AND CONSTRUCTED WITH APPROPRIATE FOUNDATIONS, ADEQUATE REINFORCEMENT, AND GOOD QUALITY MORTAR.

- STANDARD DETAILS FOR LOW FENCES OR SHORT RETAINING WALLS MAY BE AVAILABLE FROM THE LOCAL BUILDING DEPARTMENT.

Schematic Upgrade Detail A16
Approximate Cost: Depends on the design

Figure 8-3d Reproduction of FEMA 74 suggestions for securing freestanding walls or fences.

Chapter 8, Anchorage of Home Contents

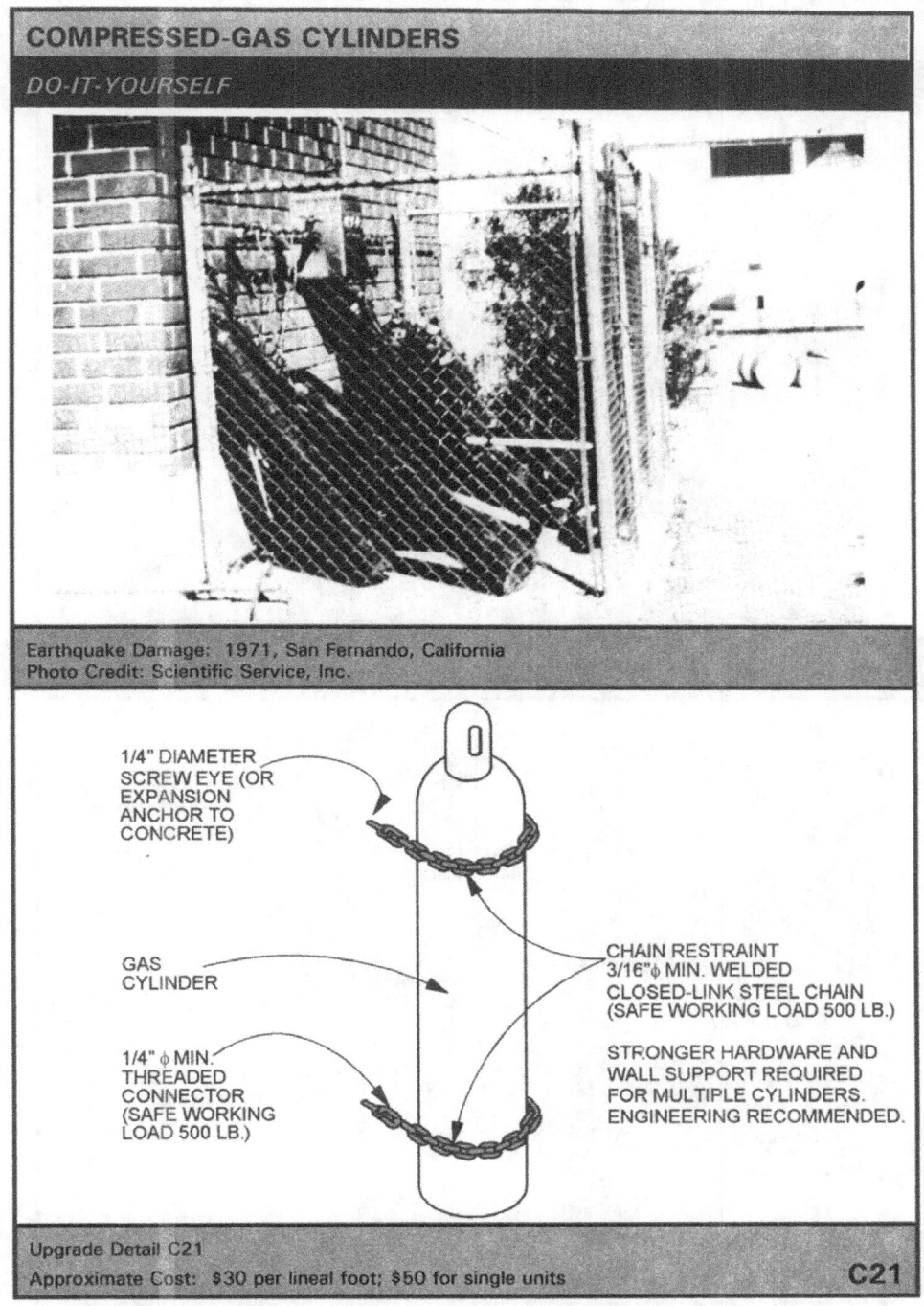

Figure 8-3e Reproduction of FEMA 74 suggestions for securing compressed-gas cylinders.

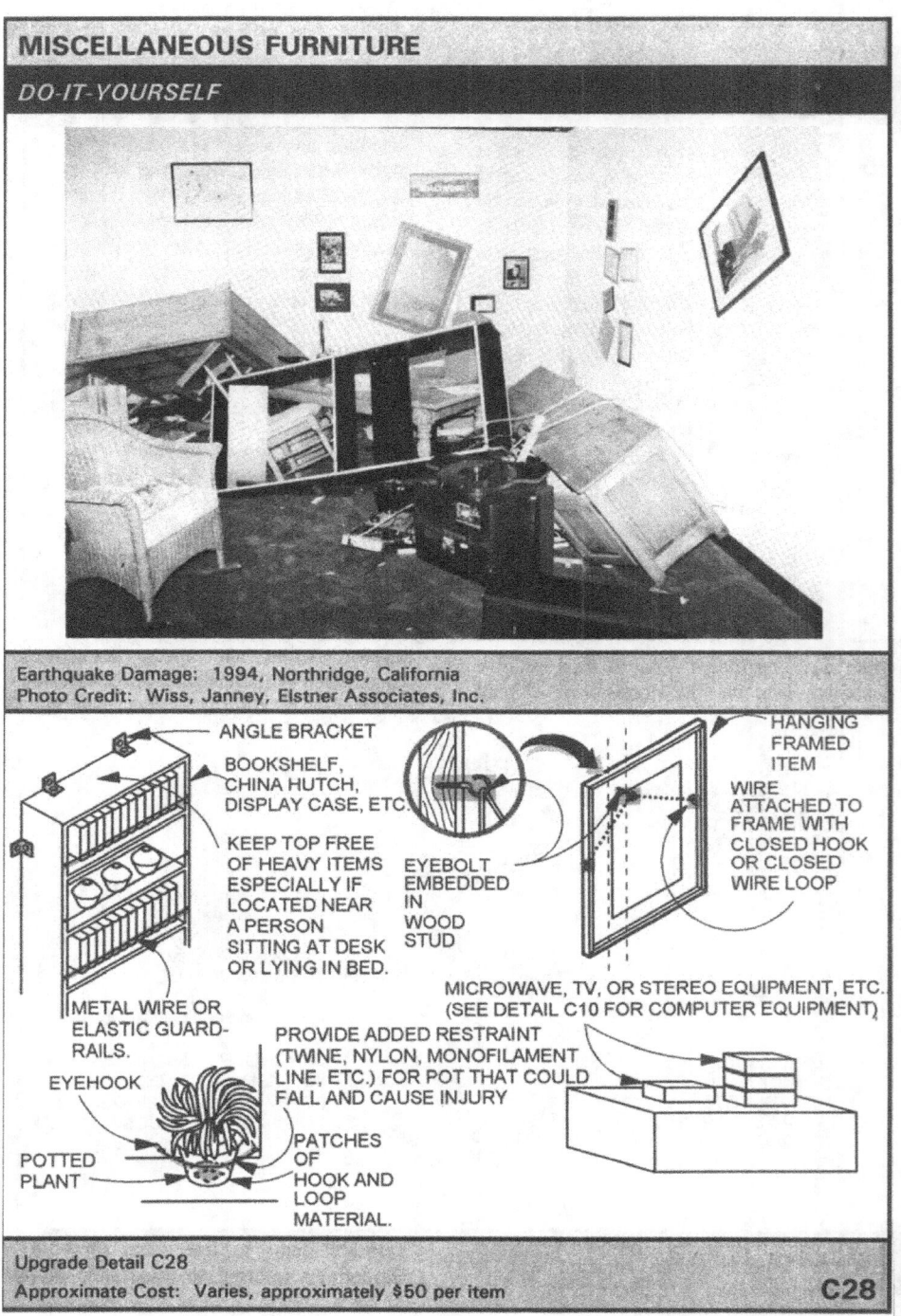

Figure 8-3f Reproduction of FEMA 74 suggestions for securing miscellaneous furniture.

Chapter 9
EXISTING HOUSES

Additions or alterations can reduce the earthquake resistance of an existing house. With proper consideration, however, earthquake resistance can be maintained or even increased as part of an addition or alteration. This chapter discusses the earthquake-resistance implications of additions and alterations and provides recommendations and references for earthquake upgrades.

9.1 ADDITIONS AND ALTERATIONS

Additions and alterations modify the load-resisting systems of existing houses. Generally, both the systems supporting gravity loads and those supporting lateral (wind and earthquake) loads are affected. For additions and alterations, *IRC* Section R102.7.1 requires that any new work conform to the *IRC*, but existing construction is allowed to remain unless it is made unsafe or will adversely affect the performance of the house. This wording provides significant opportunity for interpretation by the user and building official. *IBC* Section 3402 provides more specific guidance for acceptable reduction in strength or increase in loading, which may be appropriate to some additions and alterations. The following discussion of additions and alterations highlights issues and concerns that should be considered when interpreting *IRC* requirements.

9.1.1 Alterations

Alterations to existing houses often involve modification or removal of existing bracing walls and portions of floors and roofs. Figure 9-1 shows two alterations that remove exterior bracing walls from a house and disrupt the roof. Interior remodels often remove interior walls that provide bracing for earthquake and wind loads.

Where existing bracing walls are removed or reduced due to alterations, the remaining bracing walls should be checked for conformance with the bracing location, length, and bracing type requirements of the *IRC* provisions. The primary focus should be on bracing in the immediate vicinity of the alteration. If bracing deficiencies occur in other portions of a house, upgrade of those areas is encouraged.

When skylights, dormer windows, or similar openings are added to existing roofs, the openings should be checked for conformance with *IRC* requirements. For earthquake loading, this would include checking the opening size against permitted maximum sizes in the irregularities provisions and checking detailing against *IRC* requirements. The framing around the opening also should be checked for gravity load requirements such as doubled rafters and headers. If a significant rebuilding of the roof is occurring, a broader range of *IRC* provisions require checking as does the completeness of the load path for gravity and lateral loads.

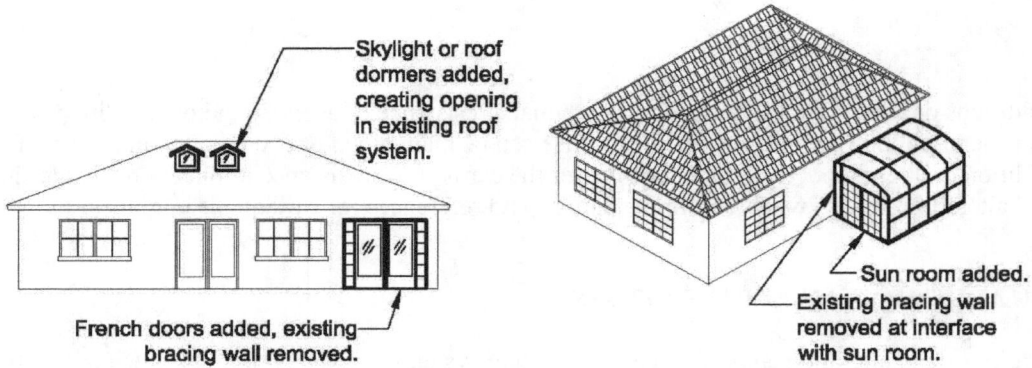

Figure 9-1 Alterations to existing house: bracing wall and roof modifications (left) and modification for the addition of sun room (right).

9.1.2 Additions

An addition to an existing house often results in both the removal of some existing bracing wall, roof, and floor areas and the addition of weight and, therefore, increased earthquake loading. Most additions can be categorized as horizontal additions, vertical additions, or a combination of the two. Horizontal additions generally are built along the side of an existing house and often require reframing of the roof. Vertical additions generally involve the addition of an upper story. Figure 9-2 illustrates horizontal and vertical additions.

Horizontal additions may create irregularities or make existing irregularities worse. Thus, the *IRC* building configuration irregularity provisions should be reviewed to assess the post-addition configuration.

Horizontal additions include the construction of new bracing walls at the new exterior of the house (and sometimes on the interior). A significant reconfiguration of bracing walls at the interface of existing and new construction also often occurs. All bracing walls in the addition and interface should be checked for conformance with the *IRC* as should any portions of the existing house where framing and bracing modifications have been made.

Chapter 9, Existing Houses

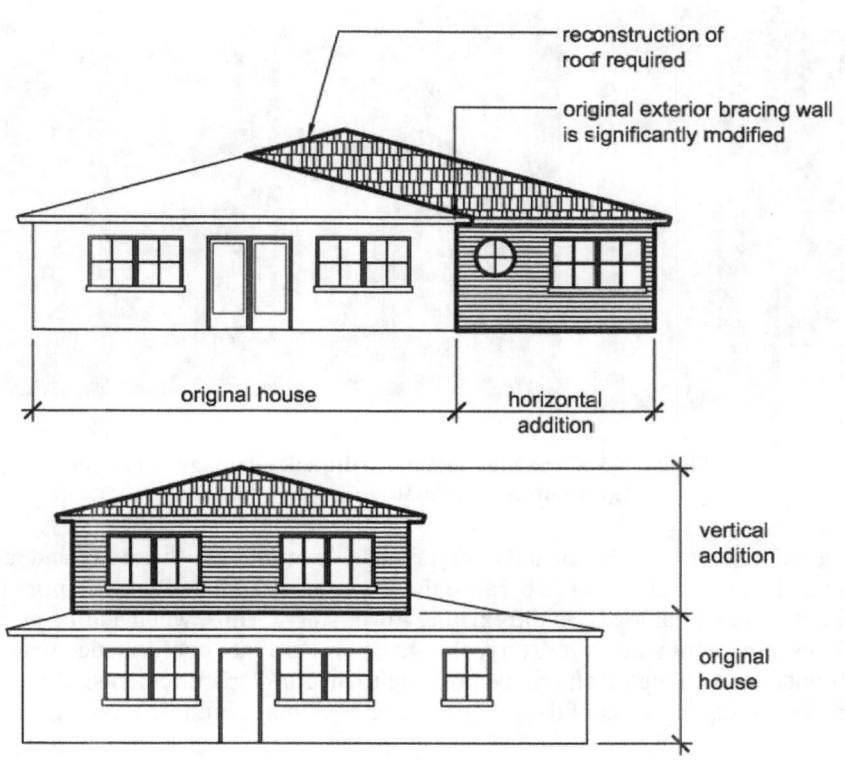

Figure 9-2 Horizontal addition (above) and vertical addition (below).

Finally, it is very important that the addition and the existing house be tied together very well. Ideally the level of interconnection should be the same as would occur if they had been built at the same time, but this generally cannot be practically achieved. Top plates and sill plates should be strapped between the new and existing construction to provide continuity.

Sheathing should be continuous and fastened to the same or interconnected framing members where possible. Where not possible, strapping of framing members should occur at a regular interval. Figure 9-3 shows the earthquake behavior of a house with failed cripple walls and adjoining (very small) slab-on-grade addition; a combination of cripple wall upgrade and strapping between the house and the addition would have greatly improved the performance of this house.

Figure 9-3 Cape Mendocino earthquake damage to cripple wall house with slab-on-grade additions.

A vertical addition demands significantly greater consideration of both gravity and lateral (earthquake and wind) loads. This is because the story that is added often will more than double both the gravity and lateral loads on the existing lower story. Thus, when adding an upper story, the entire house, including the lower story, should be brought into conformance with current code requirements. Although it should be possible to meet *IRC* requirements, use of the engineered design requirements of the *IBC* or *NFPA 5000* may result in a more practical design.

With any addition or alteration, it is important that the gravity and lateral load paths be checked in detail to ensure that they are complete and meet the requirements of the *IRC*, *IBC*, or *NFPA 5000*. Additions and alterations often create nontypical load path details, and it is important that these details result in a complete load path with load-carrying capacity that is not less than would have resulted had typical details been used. In some cases, the new detailing deviates enough from that which is typical that engineered design should be employed.

Adding to or altering an existing house offers a clear opportunity to voluntarily upgrade existing portions of the house to better resist earthquake forces. The following section describes where such upgrades might be employed.

9.2 EARTHQUAKE UPGRADE MEASURES

As noted elsewhere in this guide, the life-safety performance of houses in past earthquakes has been good with only a few exceptions. There are, however, certain conditions or portions of houses that have repeatedly resulted in earthquake damage, loss, and, in some cases, life loss or injury. Among these are:

- Missing or inadequate bolting to the foundation,
- Inadequate cripple wall bracing,

- Damage to bracing and finish materials,
- Excessive drift at garage fronts,
- Partial or complete collapse of hillside houses,
- Separation and loss of vertical support at "split-level" floor offsets, and
- Damage and collapse of masonry chimneys.

Because existing houses vary widely in configuration and construction based on age, region, siting, etc., it is necessary to identify upgrade measures appropriate to the particular house. In deciding on voluntary upgrade measures, take into account the configuration of the house and the potential benefit of the upgrade. Based on the principles discussed in this guide, some simple upgrade measures for existing houses are described below. This discussion addresses when the various upgrade measures are appropriate and suggests approximate levels of priority. Note, however, that the primary objective of most published upgrade measures for houses is reduction of hazard to life.

If an upgrade is being undertaken on a voluntary basis, the house generally will not be required to conform to all of the code requirements for new construction. The building official or authority having jurisdiction should be consulted regarding minimum requirements. It is recommended that a basis for the upgrade work (i.e., published prescriptive method or earthquake load level) be established and clearly documented. When a house is being remodeled or extensively renovated, a systematic upgrade to meet the *IRC* requirements for new construction may be reasonable and may be required by the authority having jurisdiction.

The remainder of this section provides an overview of common upgrade measures. Those interested in implementing specific measures are referred to the list of references and resources in Appendix E for further information on implementation.

9.2.1 Foundation Bolting

Many houses constructed in California before the 1950s and even later in other parts of the United States do not have anchor bolts that attach the wall bottom plate or foundation sill plate to the concrete or masonry foundation. During an earthquake, this can allow the wood framing to slide off the foundation, causing loss of vertical support and sometimes cripple wall or partial basement story collapse. Where the headroom and foundation configuration permit, this situation can be remedied by adding new anchor bolts. Common anchorage configurations are illustrated in Figure 9-4. Expansion-type anchors can be used in strong concrete and masonry foundations. Adhesive anchors are recommended for use with unreinforced masonry or weaker concrete foundations but can be used with all foundation types. When anchor bolts are added, use of steel plate washers in accordance with the *IRC* is recommended.

If there is insufficient overhead room or the foundation configuration does not permit the use of anchor bolts, a wide variety of proprietary anchors are available from manufacturers. The primary purpose of this anchorage is to transmit horizontal earthquake loads acting parallel to the foundation from the foundation sill plate to the foundation. Where used, proprietary anchors should be designed for this use and loading direction by the manufacturer.

A high priority is suggested for addition of foundation bolting to houses that are not bolted based on the relatively low upgrade cost and generally high benefit. Applicable references include the *International Residential Code* (ICC, 2003a), the *International Existing Building Code* Appendix Chapter A3 (ICC, 2003c), *Training Materials for Seismic Retrofit of Wood-Frame Homes* (ABAG), *Homeowner's Handbook* (City of San Leandro, California), and *Project Impact* (City of Seattle).

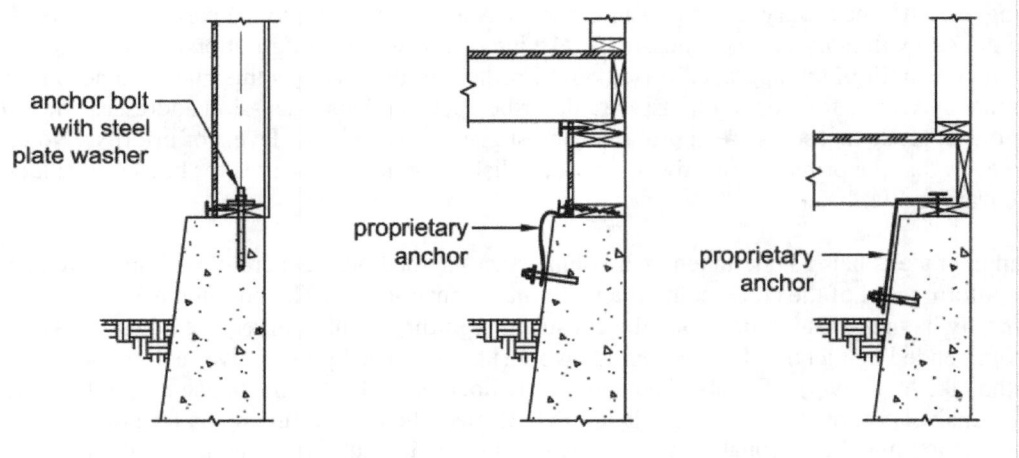

Figure 9-4 Common anchorage configurations.

9.2.2 Cripple Wall Bracing

Cripple walls are partial-height wood light-frame walls that extend from the top of the foundation to the first framed floor. Cripple walls are very susceptible to damage during an earthquake but are also one of the easiest portions of a house to upgrade for improved earthquake performance. Many of the cripple walls of existing houses (especially those constructed prior to 1960) have inadequate bracing capacity due to the type of sheathing used, inadequate attachment of the sheathing, inadequate attachment of the framing to the foundation and first floor, or decay of the system. Figure 9-5a shows a house with collapsed cripple walls (note that the porch floor is still at the original house floor elevation). As a minimum, this house will have to be jacked up to be repositioned on the foundation and have utilities reconnected. The house in Figure 9-5b suffered more severe damage due to cripple wall collapse.

Figure 9-5a House with collapsed cripple walls. Photo Courtesy of National Information Service for Earthquake Engineering, University of California, Berkeley

Figure 9-5b A house with severe damage due to cripple wall collapse. Photo Courtesy of FEMA.

Prior to an upgrade, the existing cripple wall system should be inspected and any sections of the framing that show signs of decay should be replaced. Framing materials in areas where moisture is present or in contact with the foundation should be replaced with preservative treated or decay-resistant materials. The upgrade should include anchorage of the foundation sill plate to the foundation, framing anchorage to the first-floor framing, and sheathing of the cripple walls with wood structural panel sheathing applied to either the exterior or interior face of the crawl space walls. Sheathing and connections may be installed to meet the requirements of the *IRC* or in accordance with other provisions developed specifically for upgrades. The basic elements of cripple wall bracing are illustrated in Figure 9-6.

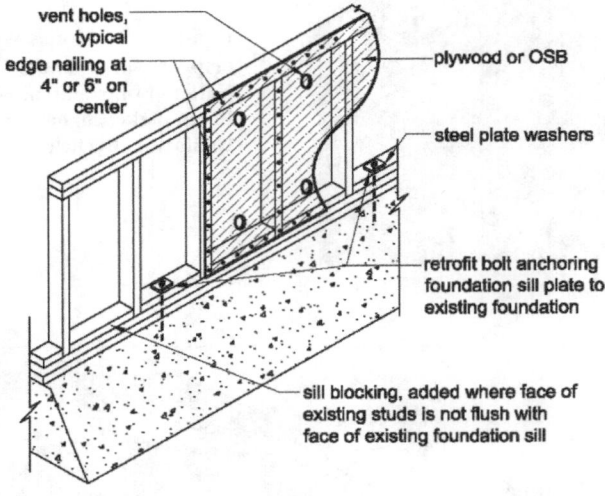

Figure 9-6 Cripple wall bracing.

Upgrade of interior cripple walls also is recommended where a crawl space is large. Where bracing is to be installed only at the perimeter of a large crawl space, performance can be improved by providing additional bracing length, reducing nail spacing from 6 inches to 4 inches, or providing sheathing on both faces of the cripple wall. A high priority is suggested for upgrading inadequately braced cripple walls based on the relatively low cost and generally high benefit. Applicable references include the *International Residential Code* (ICC, 2003a), the *International Existing Building Code* Appendix Chapter A3 (ICC, 2003c), *Training Materials For Seismic Retrofit of Wood-Frame Homes* (ABAG), *Homeowner's Handbook* (City of San Leandro), and *Project Impact* (City of Seattle).

9.2.3 Weak- and Soft-Story Bracing

Earthquake damage often is concentrated in the first story of multistory houses because the first story experiences higher loads while usually having the least amount of bracing. To reduce this damage, the first-story walls can be upgraded to increase their strength and stiffness. One method of accomplishing this is to remove the interior finish material (usually gypsum wall board or interior plaster) at the bottom of the wall in the corners of the house and to add hold-down anchors for overturning resistance. The anchors should be attached to the end studs (or other studs that have sheathing edge nailing) and the vertical rod or bolt should be attached to the foundation below (see Figure 9-7). This upgrade does not require that the entire interior finish be removed but only a section in each corner that is one stud spacing in width and several feet in height. This is an effective upgrade measure where continuous reinforced foundations exist.

Hold-down anchorage to isolated footings or unreinforced masonry footings is likely to be much less effective and engineering review is recommended.

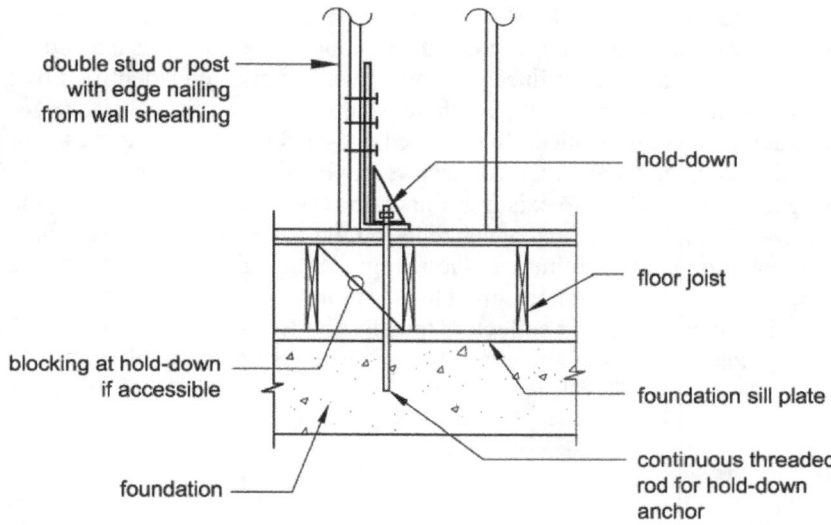

Figure 9-7 Weak- and soft-story bracing.

If the interior finish material is being removed for other reasons, then additional upgrade opportunities present themselves. These include additional anchorage of the top of the wall to the floor framing above, attachment of the bottom plate of the wall to the floor framing below, and the addition of blocking at the floor framing for the story above if not present at all locations that bear on walls. In addition, before gypsum wallboard or another finish material is attached to the walls, wood structural panel sheathing can be applied to the interior of the walls using 4-inch nail spacing around the perimeter of each sheet of sheathing and to the stud or post with the hold-down attached. A moderate priority is suggested for soft- and weak-story bracing upgrades in one- and two-family detached houses based on their moderate costs. The benefit can vary widely depending on the configuration of the existing house.

9.2.4 Open-Front Bracing

An open-front configuration occurs when bracing walls are omitted (or of grossly inadequate length) along one edge of a floor or roof. This is applicable to stories braced by light-framed walls. A number of apartment buildings with open front first stories were severely damaged or collapsed in the Loma Prieta and Northridge earthquakes. In this apartment building type, called "tuck-under parking," significant lengths of first-story bracing wall were omitted in order to provide access to under-building parking. One- and two-family houses with open-front configurations also are vulnerable to earthquake damage. Two common occurrences of open-front configurations in one- and two-family houses are the fronts of attached garages with inadequate bracing length and window walls with no bracing. These conditions are illustrated in Figure 9-8.

The open-front garage condition is of most concern when there is living space over the garage. In newer houses, narrow wall segments at the side of garage door openings may contain pre-fabricated bracing wall systems or engineered bracing walls that can be identified by the use of steel hold-down connectors or straps. Where there is no indication of such bracing systems, the installation of bracing is recommended. In detached one- and two-family houses, wood structural panel sheathing and anchorage connectors can be added in accordance with *IRC* provisions using adhesive anchors to existing foundations (Figure 9-9). The performance of narrow bracing walls with hold-down devices relies on the continuity of the existing foundation. If the existing foundation is not continuous, shows signs of damage or is constructed of unreinforced masonry or post-tensioned concrete, an engineering evaluation should be undertaken. Steel moment frames or collectors transferring loads to other portions of the house are alternative upgrade measures where use of bracing walls is not possible.

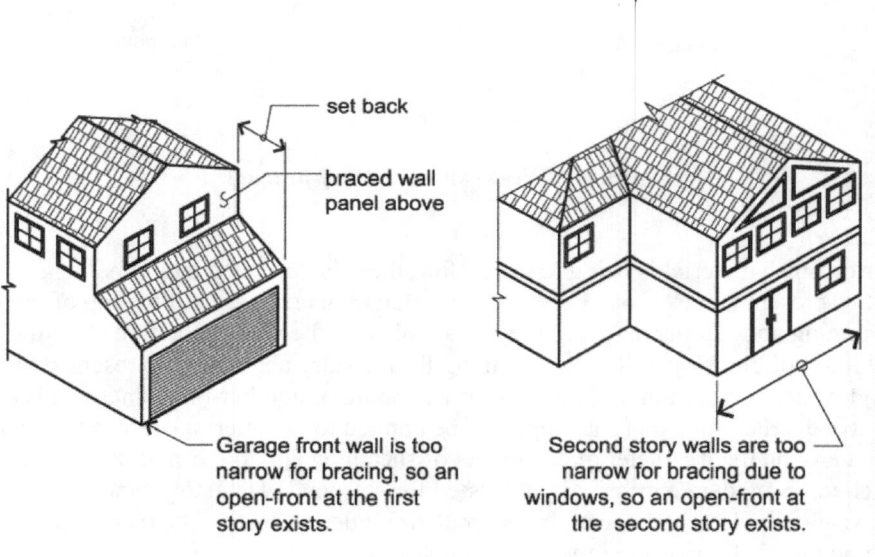

Figure 9-8 Common open-front occurrences in one- and two-family detached houses.

Chapter 9, Existing Houses

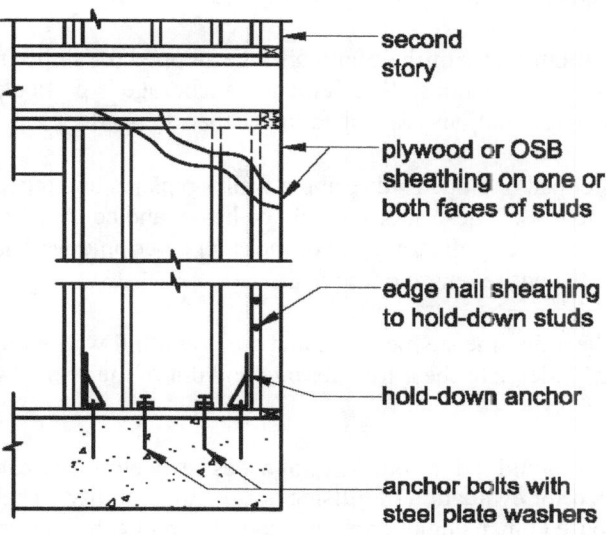

Figure 9-9 Detailing of narrow bracing wall piers at open fronts.

A high priority is suggested for upgrade of open fronts in the first stories of multistory houses. A moderate priority is suggested in single-story houses or the top story of multistory houses.

9.2.5 Hillside House Bracing

A number of houses on steep hillsides collapsed or were severely damaged in the Northridge earthquake. Where damage occurred, the lot sloped downward from the street level, and the main floor of the house was located at or near street level with either a stilt system or tall wood light-frame walls between the house and grade. Many of the failures began with floor framing pulling away from the uphill foundation or foundation wall. Typical damage to these types of houses is shown in Figure 9-10. As a result of Northridge earthquake damage, the City of Los Angeles has developed provisions for anchoring such houses to their uphill foundations to reduce the risk of failure.

There are no prescriptive upgrade measures currently available for these hillside houses so upgrade requires an engineered design. Upgrade measures to improve the response of the structure and reduce the amount of damage that occurs during an earthquake include:

- Securely anchoring the floor framing to the uphill foundation. This will require that anchors (i.e., hold-down connectors) be used to attach the floor joists to the foundation

161

using adhesive anchors (see Figure 9-11). This detail should be repeated at each framed floor level that attaches directly to an uphill foundation.

- Attaching the bottom plates of the framing of the stepped (or sloped) and downhill cripple walls to the foundations. Supplemental anchorage is particularly important where the top of the side foundations are sloped rather than stepped.

- Attaching the stepped or sloped side cripple wall top plates together at all splice joints using strap connectors. These straps should be heavy and connected securely to both sides of the splice, making the top plate of the stepped cripple wall act as though it were one piece along the entire length of the wall.

- Continuously sheathing the stepped wall and the down-hill wall with wood structural panel sheathing. Adequate shear transfer into and out of the cripple walls should be provided.

These upgrade measures should reduce but will not necessarily eliminate earthquake damage. A high priority is suggested for evaluation of hillside house vulnerability. The need for upgrade should be determined based on an engineering evaluation. Applicable references are *Voluntary Earthquake Hazard Reduction in Existing Hillside Buildings* (City of Los Angeles, 2002) and *Framing Earthquake Retrofitting Decisions: The Case of Hillside Homes in Los Angeles* (von Winterfeldt, Roselund and Kitsuse, 2000).

Figure 9-10 House located on a hillside site damaged during the Northridge, California, earthquake.

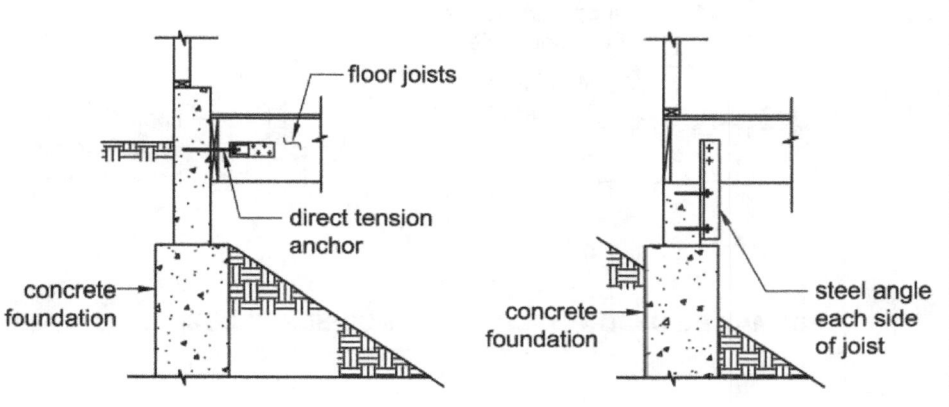

Figure 9-11 Anchorage of floor framing to the uphill foundation.

9.2.6 Split-Level Floor Interconnection

Split-level houses experienced partial collapse and significant damage in the 1971 San Fernando earthquake. These houses had vertical offsets in the floor framing elevation on either side of a common wall or other support as shown in Figure 9-12. Earthquake damage occurred when sections of floor and roof framing pulled away from the common wall. (See Section 2.3 of this guide for additional discussion of irregularities.) The behavior of split-level configurations can be improved by adequately anchoring floor framing on either side to the common wall. Where offset floors are close enough in elevation that a direct tension tie can be provided between levels, an upgrade can be accomplished with installation of steel straps; a strap spacing of not more than 8 feet on center is recommended (Figure 9-12). Where direct tension ties are not practical, ATC (1976) provides a variety of details for anchorage of framing to the supporting wall. Finish removal will often be required in order to install connections, making this upgrade most practical when remodel work is occurring. It is difficult to establish a priority for this upgrade because significant damage was observed only in the San Fernando earthquake and photos suggest that the houses damaged had little or no positive connection provided between offset floor levels. An applicable reference is *A Methodology for Seismic Design and Construction of Single-Family Dwellings* (ATC, 1976).

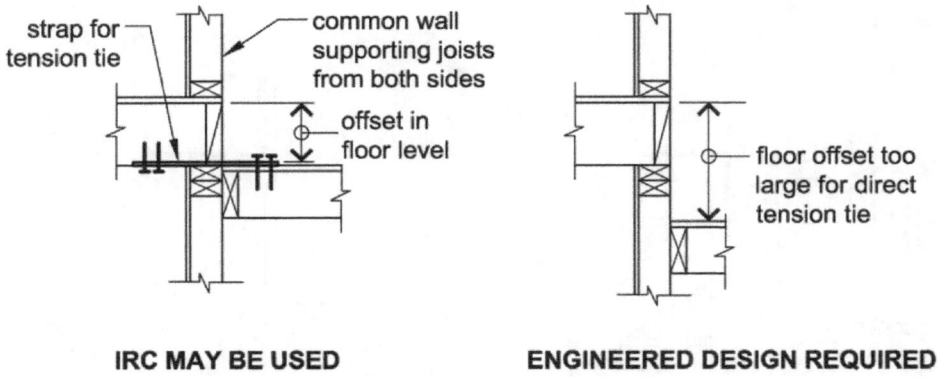

Figure 9-12 Split-level ties for floor framing.

9.2.7 Anchorage of Masonry Chimneys

Fireplaces and chimneys in new construction were discussed in Chapter 7 where it was noted that masonry and concrete fireplaces are heavy, rigid, brittle, and very susceptible to earthquake damage. *IRC* requirements for new construction in SDCs D_1 and D_2 dictate use of horizontal and vertical reinforcing and anchorage of the chimney to the framing at floors, ceilings, and roof.

Chimneys on existing houses generally are even more vulnerable than new chimneys because they seldom have reinforcing or are anchored to the house. Common chimney failures range from hairline fractures of masonry and flue liners to complete fracture (i.e., the top of the firebox and at the roof line) permitting large sections of the chimney to fall away from or into the house, shattering into a pile of rubble (see Chapter 7 for discussion and illustrations). Chimneys are reported to have caused one fatality in the 1992 Landers earthquake and critically injured one person in the 2000 Napa earthquake (Association of Bay Area Governments).

The upgrading of chimneys is very controversial within the earthquake engineering community. Upgrading of existing masonry chimneys most often includes strapping the chimney to the house at the roof, ceiling, and floor levels. Where the chimney extends a significant distance above the roof line, braces from the top of the chimney down to the roof also may be added. Advocates of chimney bracing believe that anchoring the chimney will reduce the hazard posed by falling portions of the chimney. Opponents note that even with strapping, a chimney seeing significant earthquake loading is likely to be damaged to the point that removal and reconstruction are required. Opponents further point out that upgrading does not always improve chimney performance. Both arguments deserve consideration and the reader is referred to the reference list for additional discussion. Easy and practical approaches to reducing risk to life are provided on the Association of Bay Area Governments (ABAG) website and include the suggestion that occupants should not sleep in the area immediately surrounding a fireplace with an unreinforced or unanchored chimney.

Where chimneys occur at the house exterior, steel straps similar to those discussed in Chapter 7 can be wrapped around the outside of the chimney and anchored to floor, ceiling, and roof framing in much the same way as was illustrated for new construction. Because this strap will have exterior exposure, heavy steel should be used and corrosion protection will need to be maintained. It also is important that the *IRC* Section R1001.15 gap be maintained between the chimney face and combustible framing. Although the addition of straps is not likely to keep a chimney from being damaged, it may reduce the falling hazard if it is damaged. A recommended alternative is removal of the chimney or fireplace and chimney and replacement with a factory-built fireplace and flue surrounded by light-frame walls.

Applicable references include *Info on Chimney Safety and Earthquakes* (ABAG) and *Reconstruction and Replacement of Earthquake Damaged Masonry Chimneys* (City of Los Angeles, 2001).

9.2.8 Anchorage of Concrete and Masonry Walls

Under earthquake loading, concrete and masonry walls can pull away from roof and floor framing. This is primarily a concern where bolted ledgers support framing and no direct anchorage of the wall to the framing exists. This condition can be effectively upgraded by providing a tension connection between the wall and the floor and roof framing as shown in Figure 9-13. The connection should be made to the joists when the joists run perpendicular to the wall and should be made to blocking and extend at least 4 or 5 feet into the interior of the floor system when the joists run perpendicular. An engineering evaluation of the existing condition and engineered design of upgrade measures are recommended.

Applicable references include *International Existing Building Code* Appendix Chapter A2 (ICC, 2003c) and *Guidelines for Seismic Evaluation and Rehabilitation of Tilt-up Buildings and Other Rigid Wall / Flexible Diaphragm Structures* (SEAONC, 2001).

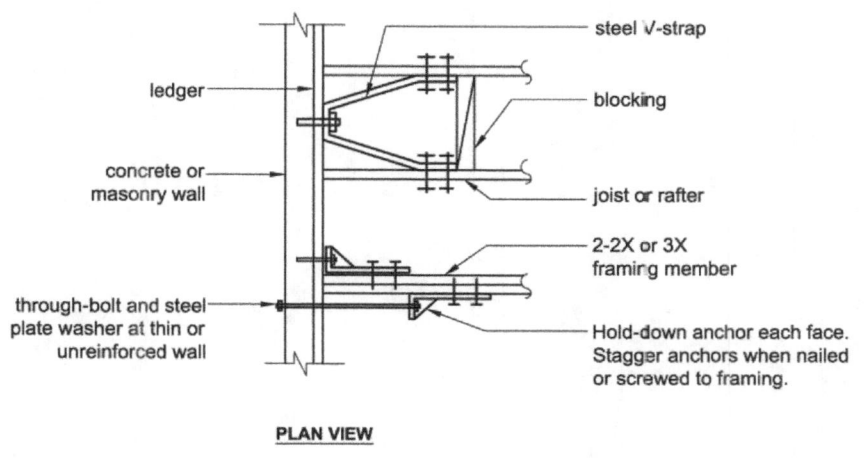

Figure 9-13 Anchorage of concrete or masonry walls to floor, roof, or ceiling framing.

Appendix A

ANALYSIS OF THE MODEL HOUSE USED IN THIS GUIDE

A model single-family detached house was developed and analyzed in preparing this guide. The house is described in Chapter 1, and the results of the analysis are referred to throughout the guide. This appendix provides additional details concerning the model house, the analysis, and the interpretation of analysis results.

The analysis of the model house provided an approximate comparison of performance for varying wood light-frame house and bracing configurations permitted by the *IRC* and permitted the assessment of improved performance resulting from application of the **above-code** recommendations made in this guide. While the model house and the analysis performed cannot represent all houses that may be constructed using *IRC* provisions, they do provide a specific example of relative performance from which trends can been observed.

A1 MODEL HOUSE

The model house contained both one-story and two-story portions, three bedrooms, 2-1/2 baths, and an area of approximately 2,500 square feet plus garage. The house design is intended to reflect current configurations for wood light-frame construction but not necessarily any specific region of the United States. Separate analytical models were developed for common variations in the design including base conditions, exterior finishes, and earthquake bracing configurations.

The base conditions are slab-on-grade construction with turned-down footings (Figure A-1), continuous exterior footings with level 2-foot-high cripple walls (Figure A-2), a hillside condition with cripple walls of varying height (Figure A-3), and a full basement with concrete or masonry walls (Figure A-4). Exterior finishes are categorized as light and veneer. The light finish is intended to represent low-weight finishes such as vinyl or fiber-cement board siding. The veneer is intended to represent a single-wythe anchored brick veneer used for the entire house exterior. Bracing requirements were determined for each configuration and Seismic Design Category (SDC) in accordance with the 2003 *IRC*. Chimneys of light-frame construction were used for all house configurations.

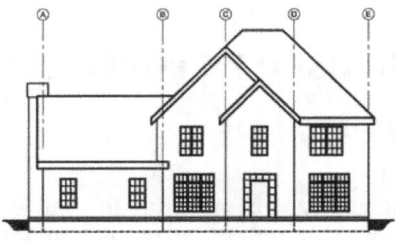

Figure A-1 Slab on grade base.

Figure A-2 Level cripple wall base.

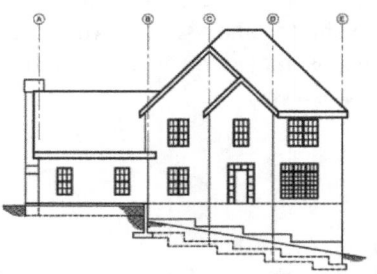

Figure A-3 Hillside base.

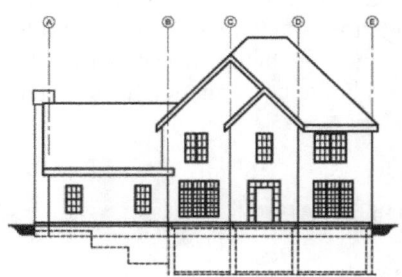

Figure A-4 Basement.

IRC prescriptive bracing requirements were determined for each combination of base condition, exterior finish, and Seismic Design Category. Because use of veneer is not permitted on houses with cripple walls in SDC D_1 and D_2 (*IRC* Section R703.7, Exceptions 3 and 4), both the level and hillside cripple wall configurations with veneer were limited to SDC C. Table A-1 is an example of one bracing spreadsheet. The remaining spreadsheets used in determining bracing requirements for each of the designs are not included here due to their length; however, they and other information used in the analysis are available upon request from the Building Seismic Safety Council.

Because gypsum wallboard is used in almost every U.S. residential building, it was used for the structural bracing wherever possible. Since it would be installed as a finish anyway, its use for bracing has the least construction cost. Wood structural panel wall bracing was used where length and percentage bracing requirements could not be met with gypsum wallboard. Alternative braced wall panels conforming to *IRC* Section R602.10.6 were used for the slender walls at the house front and garage front for slab-on-grade and basement base conditions. The alternative braced wall panels require support directly on a continuous foundation; therefore, they could not be used in combination with cripple walls. The *IRC* Section R602.10.5

"continuous structural panel sheathing" modifications to bracing length and percent were not used. See Figure 5-6a through 5-6b for an example bracing plan of a cripple wall base condition in SDC C.

Several interpretations of *IRC* requirements were made in developing the bracing designs. First, it was recognized that the roof-plus-ceiling assembly weight would fall just below the limit of 15 psf of *IRC* Section R301.2.2.2.1 if the roof assembly weight were considered based on the unit weight on slope (12:12 roof slope) but would exceed 15 psf if the weight were adjusted to horizontal projected area. Although adjustment to the horizontal projected area is common practice in engineering calculations, it was decided that this calculation is not specifically noted in the 2003 *IRC* provisions so the roof-plus-ceiling assembly weight was deemed to fall within the 15 psf *IRC* limit.

The second interpretation related to the use of gypsum wallboard bracing (*IRC* Section R602.10.3, Method 5). *IRC* Section R602.10.4 requires that gypsum wallboard braced wall panels applied to one side of a wall be at least 8 feet in width. It was interpreted to mean that a continuous length of full-height wall not less than 8 feet wide would have to be available in order to use this bracing method. Interruption of the 8-foot length by perpendicular walls was interpreted to mean that it was not permitted. Where an 8-foot length of full-height wall was not available, wood structural panels were used as bracing instead. Based on this interpretation and the configuration of the model house, wood structural panels rather than gypsum wallboard were used for a significant portion of the exterior wall bracing. Where gypsum wallboard bracing can be applied to both faces of a wall (such as at interior walls), the minimum required length of full-height sheathing is reduced to 4 feet. While the perforated shear wall method that includes hold-down anchorage at the ends of the wall line was used as an **above-code** option for the analysis, the continuous sheathed option of *IRC* Section R602.10.5 that allows a 10 percent reduction in the sheathing percentage was not used in the analysis.

The third interpretation relates to the bracing requirements used for the model house in SDC C. *IRC* Table R602.10.1 specifically identifies sheathing length requirements for SDC C. Some *IRC* users, however, interpret the *IRC* Section R301.2.2 exception to mean that the table bracing length requirements for SDCs A and B can be used for houses in SDC C. Analysis of the model house performed for this guide used the SDC C bracing length requirements.

The resulting bracing configurations are illustrated on a set of bracing plans and elevations for each of the designs are available from the Building Seismic Safety Council on a CD-ROM. The increased bracing length requirements for higher Seismic Design Categories can be observed to have reduced allowable door and window openings.

Appendix A, Analysis of the Model House Used in This Guide

Table A-1 Example Wall Bracing per 2003 *IRC*, Slab-on-grade Base Condition

Seismic Design Category	Wall Finish	Total Stories	Story Considered	Wall Line	Wall Line Length (ft)	Type 3 Percent	Type 3 Length (ft)	Other Type Percent	Other Type Length (ft)	Adjustments					Type 3 Adjusted Length (ft)	Other Type Adjusted Length (ft)
										Wall Line Spacing R602.10.1.1	Wall Wt. Table R602.10.3 Footnote d (1)	Roof + Ceil. Table R301.2.2.4	Veneer R703.7 Exc. 2		
C	light	2	2	B	33.5	16	5.4	25	8.4	1.06	0.85	1.00	1.00	4.8	7.5
			2	E	30	16	4.8	25	7.5	1.06	1.00	1.00	1.00	5.1	7.9
			2	1,7	37	16	5.9	25	9.3	1.00	1.00	1.00	1.00	5.9	9.3
			2	5, 6	37	16	5.9	25	9.3	1.00	1.00	1.00	1.00	5.9	9.3
		1 (2)	1	A	40	16	6.4	25	10.0	1.00	1.00	1.00	1.00	6.4	10.0
			1	1	20	16	3.2	25	5.0	1.14	1.00	1.00	1.00	3.7	5.7
			1	8	20	16	3.2	25	5.0	1.14	1.00	1.00	1.00	3.7	5.7
			1	B	40	30	12.0	45	18.0	1.06	0.85	1.00	1.00	10.8	16.2
			1	E	29	30	8.7	45	13.1	1.06	0.85	1.00	1.00	7.8	11.7
		2	1	2	37	30	11.1	45	16.7	1.00	0.85	1.00	1.00	9.4	14.2
			1	5, 6	37	30	11.1	45	16.7	1.00	0.85	1.00	1.00	9.4	14.2
C	veneer	2	2	B	33.5	16	5.4	25	8.4	1.06	1.00	1.00	1.00	5.7	8.9
			2	E	30	16	4.8	25	7.5	1.06	1.00	1.00	1.00	5.1	7.9
			2	1,7	37	16	5.9	25	9.3	1.00	1.00	1.00	1.00	5.9	9.3
			2	5, 6	37	16	5.9	25	9.3	1.00	1.00	1.00	1.00	5.9	9.3
		1 (2)	1	A	40	16	6.4	25	10.0	1.00	1.00	1.00	1.00	6.4	10.0
			1	1	20	16	3.2	25	5.0	1.14	1.00	1.00	1.00	3.7	5.7
			1	8	20	16	3.2	25	5.0	1.14	1.00	1.00	1.00	3.7	5.7
			1	B	40	30	12.0	45	18.0	1.06	1.00	1.00	1.50	19.0	28.5
			1	E	29	30	8.7	45	13.1	1.06	1.00	1.00	1.50	13.8	20.7
		2	1	2	37	30	11.1	45	16.7	1.00	1.00	1.00	1.50	16.7	25.0
			1	5, 6, 7	37	30	11.1	45	16.7	1.00	1.00	1.00	1.50	16.7	25.0

[1] Reduction cannot be applied to top-most story where resulting bracing length would be less than required for wind.
[2] The garage and family room areas are treated as a one-story building attached to the two-story house.

A2 ANALYSIS USING STANDARD ENGINEERED DESIGN METHODS

Prior to evaluation using other methods, earthquake forces and deformations were estimated using the linear static methods commonly used in engineering design of new buildings. Included were force calculations using the *International Building Code* (*IBC*) linear static method, estimation of drift using the APA-The Engineered Wood Association four-term shear wall deflection equations at strength level forces, and amplification to estimated drifts using *IBC* amplification factors. This approach resulted in the APA shear wall deflection equations being used outside of their intended range (based on force per nail limits included with nail slip variables). This provided clearly unrealistic shear wall deflections amplified to unrealistic estimated drifts (over 36 inch drifts in some cases). Thus, it was concluded that the use of these engineered design estimates as predictors of performance for non-engineered buildings was not realistic and it was not pursued. Likewise, use of other available deflection equations that represent simplifications of the APA equations were not pursued. This issue should not occur when using this standard deflection calculation method for engineered buildings.

A3 ANALYSIS USING NONLINEAR METHODS

Nonlinear time-history analysis using the Seismic Analysis of Woodframe Structures (SAWS) analysis program was chosen as the best available method for estimating force and deformation demands based on analytical studies that were verified against shake table results from the FEMA-funded CUREE-Caltech Woodframe Project. Analysis models included both designated bracing and finish materials. The Woodframe Project analytically predicted forces and deflections compared favorably with shake-table results and were clearly differentiated from analysis and testing results without finish materials (Folz and Filiatrault, 2002).

The SAWS analysis program uses rigid diaphragms to represent floor and roof diaphragms. Walls are modeled as nonlinear springs with hysteretic parameters developed specifically to describe the behavior of woodframe bracing systems. For the example house, rigid diaphragms were used to represent the high roof, the low roof plus second floor, and, where appropriate, the first floor. A simplified representation of the rigid diaphragms and wall springs for the model house is presented in Figure A-5.

Ten sets of hysteretic parameters were developed from component testing data to describe wall bracing and interior gypsum wallboard finishes. Figure A-6 illustrates the meaning of the parameters, and a summary of analytical modeling parameter values is provided in Table A-2. For each of the bracing materials (with the exception of No. 5 and 6), the hysteretic parameters were determined for a 4-foot bracing length. Because widely varying lengths are used in the house, the parameters were scaled for varying bracing lengths.

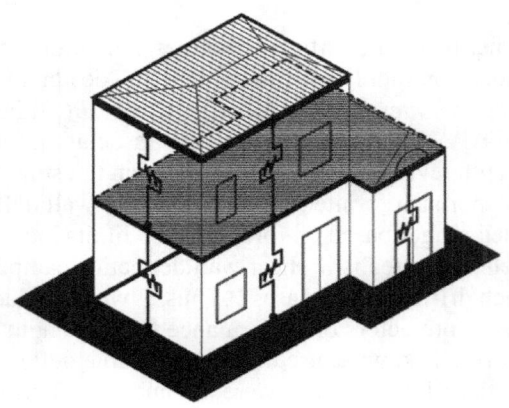

Figure A-5 Analysis model.

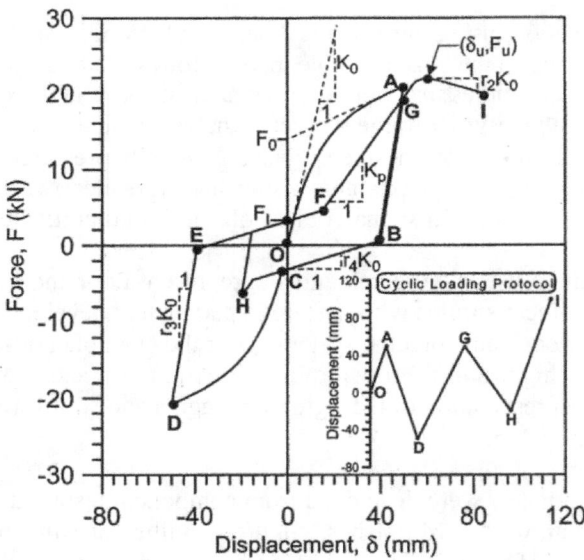

Figure A-6 Hysteretic parameters for model ((Folz and Filiatrault, 2002).

Hysteretic parameters currently available from laboratory testing of wall components vary based on wall boundary conditions, test set-up, and test protocol. Parameters chosen for the analysis of the model house tended towards lower bounds of strength and stiffness. Future analytical studies should consider exploring upper and lower bounds.

In order to simplify interpretation of analysis results, the analysis model uses consistent identification of bracing walls across all building configurations. Because of this modeling approach, cripple walls have been included in the model for all building configurations; where slab-on-grade construction occurs, the cripple walls are modeled as extremely rigid elements (Property No. 9) resulting in negligible deflection. In addition, some wall elements occur only in limited configurations; a bracing length of 0.1 foot is used where a bracing panel is intended to have no effect.

Table A-2 Hysteretic Parameters Used for Nonlinear Dynamic Model

Property No.	Use	Fo (k)	Fl (k)	Δ_u (in)	So (Ko) (k/in)	r1	r2	r3	r4	alpha	beta	Δ_{CUREE} (in)	Fu
					Basic Model Properties								
3	gypsum wallboard lower bound[1]	0.50	0.20	0.70	3.0	0.130	-0.050	1.000	0.030	0.40	1.10	0.80	0.47
4	4 ft OSB w/o tie-down	0.59	0.15	1.00	2.8	0.096	-0.021	1.000	0.010	0.40	1.10	1.90	0.63
5	2'-8" wall one-sided	1.20	0.20	2.70	3.0	0.020	-0.100	1.000	0.024	0.87	1.10	2.00	0.62
6	2'-8" wall two-sided	2.40	0.40	2.70	6.0	0.020	-0.100	1.000	0.024	0.87	1.10	2.00	1.24
7	2 ft high cripple wall[1]	3.00	0.30	1.00	20.0	0.010	-0.050	1.000	0.010	0.87	1.10	1.00	3.20
8	1:3 stepped cripple wall[1]	3.30	0.60	1.00	23.0	0.090	-0.110	1.000	0.040	0.87	1.10	0.90	5.48
9	zero height cripple wall = 100 x No. 7	300.00	30.00	1.00	2000.0	0.010	-0.050	1.000	0.010	0.87	1.10	1.00	320.00
					Code Plus Properties								
10	Code Plus fully sheathed perforated shear wall[1]	2.30	0.05	2.20	2.6	0.200	-1.000	1.000	0.020	0.87	1.10	1.50	3.00
11	4 ft OSB w/ tie-down	1.70	0.50	3.00	6.8	0.040	-0.056	1.000	0.024	0.87	1.10	2.60	2.51
12													

[1]Values in table are for 4 ft length. Fo, Fi & Ko in model are scaled by length of full height sheathing. See sepatate spreadsheets.

The strength and stiffness contribution of exterior wall finishes was not included in the analysis. This approach was chosen because it would lead to a lower bound and, therefore, conservative estimate of deformation demand. In addition, some exterior finish materials are believed to have very little impact on building behavior (e.g., vinyl siding) and information was not available on the contribution of some other finishes (e.g., brick veneer). Due to the judgment necessary to select appropriate component testing and to derive parameters and the simplification of not including exterior finishes, the resulting modeling must be qualified as being approximate.

Earthquake demand is represented using the larger horizontal acceleration record from Canoga Park for the 1994 Northridge, California, earthquake. This record was chosen because it corresponds well with the code design spectra over a range of building periods. The peak acceleration was scaled for each Seismic Design Category by dividing the maximum S_{DS} value for each category by 2.5, resulting in peak accelerations of 0.2, 0.33 and 0.47g for SDCs C, D_1, and D_2, respectively. For comparison, the recorded ground motion has a peak acceleration of

0.42g and was scaled to 0.50g to represent Zone 4 anticipated ground motions in the CUREE shake-table testing. The ground motion scaling used for this analysis represents the demand used as a basis for code design. The demand from the maximum considered earthquake (MCE) ground motion (MCE) would be approximately 50 percent greater.

Detailed assembly weights and building weights have been determined for each house configuration. The analysis model spreads the resulting mass uniformly over a single rectangle used to describe each above-ground diaphragm. The center of the mass rectangle is set at the calculated center of mass of the building. This simplification, made necessary by analysis limitations, should have a minor effect on results.

A4 ANALYSIS RESULTS

The selected ground motion was run once in the horizontal X-direction and once in the horizontal Y-direction for each combination of base condition, exterior finish, and Seismic Design Category as well as for a series of **above-code** recommendations. From the nonlinear time-history analysis, peak drifts in each of the bracing wall lines and peak reactions to supporting foundations were extracted and summarized in tables. These tables are not included here for brevity but are available upon request from the Building Seismic Safety Council on the analysis CD. The "controlling" value was the largest absolute value of the X- and Y-directions.

A4.1 Deformation Demand Relation to Performance

In order to translate the results of the analysis into an approximation of house performance, three ranges of peak transient wall drift and associated approximate descriptions of building performance were developed. The choice of range and description of performance are based on component and full-building test results combined with the opinions of those participating in the development of this guide.

The approximate performance categories and corresponding drift ranges are:

- **Minor** damage potential – Less than or equal to 0.5% story drift

    The house is assumed to suffer minor nonstructural damage such as cracked plaster or gypsum wallboard and hopefully would be "green-tagged" (occupancy not limited) by inspectors after an earthquake, which would permit immediate occupancy. Some repairs should still be anticipated.

- **Moderate** damage potential – Above 0.5% to 1.5% story drift

    The house is assumed to suffer moderate damage including possible significant damage to materials and associated structural damage, but the building is assumed to have some reserve capacity in terms of strength and displacement capacity. The house hopefully would be "green-tagged" or, more likely, "yellow-tagged" (limited occupancy) by inspectors after an earthquake and may or may not be habitable. Significant repairs should be anticipated.

- **Significant** damage potential – Greater than 1.5% story drift

    The house is assumed to have significant structural and nonstructural damage that could result in its being "red-tagged" (occupancy prohibited) by inspectors after an earthquake. Significant repairs to most components of the building should be anticipated, and it may be more economical to replace the house rather than repair it.

Use of these three categories permits an approximate comparison of the relative performance of different *IRC* bracing solutions and **above-code** recommendations.

A4.2 Discussion of Results

Selected results of peak drift values and approximate performance category are provided in Tables A-3 and A-4. In most cases, the drift increased with increased SDC in spite of the bracing requirements also having increased. The approximate performance often increased from minor or moderate to significant as the SDC went from C to D_2. The primary reason is the inclusion of interior gypsum wallboard in the models for all Seismic Design Categories. As the SDC increased, interior walls became required braced wall panels per *IRC* requirements rather than simply nonstructural partition walls; however, the analytical model did not change because the interior walls had already been included. The result was application of a higher demand to a model with only nominal increases in resistance.

Table A-3 Selected Results for *IRC* Bracing Provisions, Slab-on-grade Base Condition

Walls	Seismic Design Category	1st Story Peak Drift (in.) and Approximate Performance
Light	C	0.46 **Minor**
Light	D_1	1.02 **Moderate**
Light	D_2	1.72 **Significant**
Veneer	C	1.29 **Moderate**
Veneer	D_1	1.34 **Moderate**
Veneer	D_2	2.21 **Significant**

Although the building mass increased significantly with the addition of brick veneer, the increase in drift ranged from moderate to slight. This is due to the *IRC* requirement for wood structural panel sheathing and hold-down devices for veneer in SDCs D_1 and D_2. The analysis model differentiated between wood structural panel shear walls with and without hold-down devices so the different strength, stiffness, and deformation capacity were accounted for. Because of this, the *IRC* bracing required for brick veneer was seen to partially compensate for the increased demand.

The **above-code** measures were applied to the slab-on-grade base condition. The measures were seen to generally reduce the building drift, although drift increases were seen in a few walls due to changes in diaphragm rotation. In SDC D_2, the approximate performance was improved by one category for all three **above-code** measures. In SDCs C and D_1, significant decreases in drift were seen within an approximate performance category.

Table A-4 Selected Results for Above-code Measures, Slab-on-grade Base Condition

Above-code Recommendation	Walls	1st Story Peak Drift (in.), Approximate Performance, and Maximum 1st Story Drift Reduction		
		SDC C	SDC D_1	SDC D_2
Original Code Minimum	Light	0.5 **Minor** Damage	1.0 **Moderate** Damage	1.7 **Significant** Damage
Above-code Continuous Sheathing	Light	0.3 **Minor** Damage 39 percent	0.7 **Moderate** Damage 42 percent	1.2 **Moderate** Damage 30 percent
Above-code Hold-downs	Light	0.5 **Minor** Damage 28 percent	0.9 **Moderate** Damage 36 percent	1.0 **Moderate** Damage 47 percent
Above-code Lap on Rim Joist	Light	0.5 **Minor** Damage 6 percent	1.0 **Moderate** Damage 7 percent	1.4 **Moderate** Damage 24 percent

The cost of implementing each **above-code** measure during construction of the house was estimated in terms of percentage change to the construction cost for the basic house structure. Comparison to total house cost was not made because variations in finishes and fixtures can dramatically vary the house cost.

Use of continuous wood structural panel wall sheathing (fully sheathed) with overturning anchors in the corners of the house significantly reduced the drift in all Seismic Design Categories, and the approximate performance category was increased by one in SDC D_2. The cost of making this change was estimated to be 9 to 10 percent of the cost of the structural portion of the model house used in this guide.

The addition of hold-down anchors at the ends of each full-height wall segment (at the corners and edges of each door and window) significantly reduced the drift in all Seismic Design

Categories, and the approximate performance category was increased by one step in SDC D_2. For the model house, the cost of implementing this improvement was estimated to be 18 percent of the structural cost of the house.

Lapping wood structural panel wall sheathing over the band joist of the floors did not have a significant effect in SDC C or D_1 but did improve the approximate performance category by one in SDC D_2. The cost of implementing this improvement was estimated to be 0.5 percent of the cost of the structural portion of the house. This **above-code** measure can be accomplished by either sheathing the wall with oversized panels (9-foot panels on an 8-foot wall) or cutting and blocking standard size sheets.

Use of the **above-code** measures in combination is thought to have a cumulative effect in improving performance and so is encouraged.

Appendix B
EARTHQUAKE PROVISIONS CHECKLIST FOR BUILDERS AND DESIGNERS

General Earthquake-Resistance Requirements

Load Path

C   NC   N/A   Priority

☐ ☐ ☐ **HIGH** Foundation anchor bolts. *IRC* Section R403.1.6 for all SDCs. *IRC* Section R403.1.6.1 and R602.11.1 for foundation anchor bolts and plate washers in SDCs D_1 and D_2 (and townhouses in SDC C).

Typical 1/2-inch diameter bolts at 6 feet along all exterior walls. Also, 1/2-inch bolts at 6 feet along interior braced walls and interior bearing walls supported on a continuous foundation, in SDCs D_1 and D_2 (and townhouses in SDC C); 1/2-inch at 4 feet along all exterior walls and along all interior bearing walls and interior bracing walls supported on a continuous foundation for three-story in SDC D_1 (and three-story townhouses in SDC C).

3x3x1/4-inch steel plate washers: On all required anchor bolts in SDCs D_1 and D_2 (and townhouses in SDC C).

First anchor bolt should be placed seven bolt diameters minimum (3-1/2 inches for 1/2-inch diameter, 4-1/2 inches for 5/8-inch diameter) and 12 inches maximum from each end of a foundation sill plate.

Sheathing and framing fasteners (except for 1/2-diameter or larger steel bolts) used in pressure preservatively treated wood framing members must have corrosion-resistant coating or be of corrosion-resistant material.

☐ ☐ ☐ **HIGH** Overturning Anchorage. *IRC* Section R602.10.6 for alternate braced wall panels, all SDCs. *IRC* Section R602.10.11, second paragraph, Exception 2, where braced wall panels are not located at corners in SDCs D_1 and D_2. *IRC* Section R703.7, Exceptions 3 and 4, when veneer is used in SDCs D_1 and D_2. Overturning load path to foundation needed where hold-down anchors are used.

Load Path Above-code Recommendations:

- **Provide 4-foot anchor bolt spacing along all exterior and interior braced wall lines for two-story houses in SDCs D_1 and D_2.**

- **Provide continuous foundation below interior braced walls with anchor bolts at spacing of 6 feet of less in all SDCs.**

- **Provide corrosion-resistant coatings for anchor bolts installed through pressure preservatively treated foundation sill plates in all SDCs.**

- **Do not "wet set" anchor bolts; securely place anchor bolts prior to placing concrete.**

- **In SDC C, provide overturning anchorage as required in SDCs D_1 and D_2 for braced wall panels not located at corners and for houses with masonry veneer.**

- **Add tie straps between first and second story corner studs to tie the walls together in SDCs C, D_1 and D_2.**

- **Use oversized sheathing panels on exterior walls and lap over rim-joist. Nail both into the plates (top and bottom) and the rim-joists in all SDCs.**

☐ ☐ ☐   **MED** Minimum fastening. All SDCs per *IRC* Table R602.3.

☐ ☐ ☐   **MED** Designed collector members aligned with and connected to the top plate of braced walls (continuous from the end of a braced wall line to the end of the braced wall panel closest to the end of the wall line). In all SDCs where the first braced wall panel begins more than 12 feet from the end of a braced wall line. (Section *IRC* R602.10.1)

Also, in SDCs D_1 and D_2, where the braced wall panel uses wood structural panel sheathing and is not located at the end of the wall line. However, when either (a) a minimum 24-inch-wide panel is provided each side of the wall corner or (b) the braced panel end closest to the corner is provide with a hold-down, a designed collector is not required if the wood structural panel braced wall panel is located 8 feet or less from the end of the braced wall line. (*IRC* Section R602.10.11 last paragraph)

Foundations and Foundation Walls

Concrete Foundations

C   NC   N/A   Priority

☐   ☐   ☐   **HIGH** Horizontal reinforcing. In SDCs D_1 and D_2, typical one No.4 in footing. Additional No. 4 in concrete stem wall (if stem wall occurs). One No. 4 top and bottom in thickened slab footing, with alternate of one No. 5 or two No. 4 in middle third of footing height for thickened slab footings cast monolithically with slab. (*IRC* Sections R403.1.3 through R403.1.3.2)

☐   ☐   ☐   **HIGH** Vertical reinforcing. In SDCs D_1 and D_2, one No. 4 at 48 inches maximum spacing where a pour joint occurs between concrete footing and concrete stem wall. (*IRC* Section R403.1.3)

☐   ☐   ☐   **MED** Adequate support of reinforcing and anchor bolts. Reinforcing concrete cover distances of 3 inches when cast against earth and 1-1/2 inches when concrete will be exposed to weather.

☐   ☐   ☐   **HIGH** Clean footing excavations before casting concrete. Proper concrete consolidation. No water added to concrete mix at site.

☐   ☐   ☐   **MED** Minimum concrete strength. 2500 psi, all SDCs. 3000 or 3500 psi in moderate or severe weathering probability areas (*IRC* Section R402.2 and Table R402.2).

☐   ☐   ☐   **MED** Rebar lap splice length of 24-inches (straight lap) (*IRC* Section R611.7.1.2). Rebar bend radius (outer) of 2 inches for No. 4 and 2-1/2 inches for No. 5. Hook at corners and intersections of 8 inches for No. 4 and 10 inches for No. 5.

Masonry Foundations

C   NC   N/A

☐   ☐   ☐   **HIGH** Horizontal reinforcing.

☐   ☐   ☐   **HIGH** Vertical reinforcing.

☐   ☐   ☐   **MED** Rebar lap splice length of 24-inch (straight lap) (*IRC* Section R606.11.2.2.3). Rebar bend radius (outer) of 2 inches for No. 4 and 2-1/2 inches for No. 5. Hook at corners and intersections of 8 inches for No. 4 and 10 inches for No. 5 (*IRC* Section R606.11.7.4).

Foundation Walls

C   NC   N/A

☐   ☐   ☐   **HIGH** Wall thickness.

☐   ☐   ☐   **HIGH** Horizontal reinforcing.

☐   ☐   ☐   **HIGH** Vertical reinforcing.

☐ ☐ ☐ **MED** Rebar lap splice length of 24-inches (straight lap). Rebar bend radius (outer) of 2 inches for No. 4 and 2-1/2 inches for No. 5. Hook at corners and intersections of 8 inches for No. 4 and 10 inches for No. 5

Above-code Recommendations:

- Avoid construction of slab on grade homes on cut and fill sites where possible. Where this condition cannot be avoided, provide additional quality control for fill placement and compaction operations.

- Regardless of SDC, provide not less than one continuous horizontal No. 4 reinforcing bar in concrete footings. Provide a second No. 4 horizontal bar in stem wall if occurs. This will provide tension and bending capacity to help mitigate foundation damage due to earthquake, wind, soil movement and frost heave.

- Regardless of SDC, remove lose debris in the construction joint between a concrete footing and a separately cast slab-on-grade.

- In SDCs C, D1 and D2, provide not less than No. 4 at 4-feet vertical bars as dowels between a concrete footing and a separately cast slab on grade.

- Regardless of SDC, provide not less than one continuous No. 4 reinforcing bar in masonry foundations.

Floor Construction

C NC N/A

☐ ☐ ☐ **HIGH** Floor sheathing nailing. Floor sheathing should be edge nailed to blocking above all braced wall lines, exterior and interior, as part of the load path (*IRC* Table R602.3(1), Footnote i). Blocking with edge nailing needs to have a load path to top of braced wall panels.

☐ ☐ ☐ **MED** SDCs D_1 and D_2, blocking or lateral restraint. Required at intermediate floor framing member supports. (*IRC* Section R502.7, Exception)

Light-Frame Wall Construction

☐ ☐ ☐ **HIGH** Overdriven sheathing nails. For wood structural panel sheathing, nails are to be driven so that the top of the head is flush with the face of the sheathing. (It is recommended that where nail heads occasionally are more than 1/16-inch below the surface, an additional nail should be provided between existing nails. If a substantial number of nails are overdriven, the sheathing should be removed and the framing checked for splitting before replacing the sheathing with proper nails.)

☐ ☐ ☐ **HIGH** Sheathing nailing to hold-down posts and studs. Hold-downs cannot carry any load unless the wall sheathing is edge-nailed to the hold-down post or stud.

☐ ☐ ☐ **HIGH** Threaded rods with properly attached nuts need to be in place before the wall sheathing is attached to the second side of the walls.

Above-code Recommendations:

- **Increase first-story strength and stiffness to mitigate weak-story irregularity. Approaches include: (a) use of wood structural panel wall bracing and hold-down connectors at each end of each full height wall segment, (b) fully sheathing all exterior walls including below windows and above and below doors and providing hold-down connectors at building corners, and (c) providing more than the minimum braced wall panel length.**

- **Increase cripple wall strength and stiffness to mitigate weak-story irregularity by sheathing full length of exterior cripple walls.**

- **Use oversized sheathing panels on exterior walls to increase wall stiffness and strength. Lap the sheathing over the floor joists and nail to both the plates (top and bottom) and the floor joists.**

Roof Construction

C NC N/A

☐ ☐ ☐ **HIGH** Sheathing nailing at braced wall lines. Roof sheathing should be edge nailed (to blocking where present) above all braced wall lines, exterior and interior, as part of the load path (*IRC* Table R602.3(1), Footnote i). Blocking with edge nailing needs to be nailed to the top of the braced wall below to provide a complete load path.

Cold-formed Steel Construction

C NC N/A

☐ ☐ ☐ **HIGH** Load path connections. Connection of cold-formed steel framing members is different than wood light-frame connection. The *IRC* provisions include a significant number of specific connection details. Attention to these details is important to the building performance for all load types.

Above-code Recommendations:

- Add interior cold-formed steel braced walls such that the distance between braced wall lines does not exceed 35 feet.

- In all SDCs, apply the irregularity limitations developed for wood light-frame houses (*IRC* Section R301.2.2.2.2).

Masonry Wall Buildings

C   NC   N/A

☐   ☐   ☐   **HIGH** Construction quality control. Proper type of mortar for masonry being used and proper mortar mixing. (Type N mortar is prohibited in higher SDCs.) Proper placement of reinforcing. Adequate support and attachment of reinforcing and anchor bolts. Cleaning out of grout space to allow proper grout placement, including cleaning out excess mortar if necessary. Provide cleanouts if necessary for adequate cleaning. Consolidation of grout.

Above-code Recommendations:

- Each exterior wall and each interior braced wall should have one, and preferably two, sections of solid wall not less than 4 feet in length.

- Sections of solid wall should be spaced no more than 40 feet on center and should be placed as symmetrically as possible.

- All masonry walls should be supported on substantial continuous footings extended to a depth that provides competent bearing.

- The distribution of interior masonry braced walls should be carefully balanced and floor and roof plans should use simple rectangular shapes without jogs and openings.

- Apply the irregularity limitations developed for wood light-frame houses (*IRC* Section R 301.2.2.2.2). The *IRC* exceptions to Irregularities 2 and 5 can be applied but the rest of the exceptions are not applicable.

- Solid portions of wall should be stocked from floor to floor and masonry walls should be continuous from the top of the structure to the foundation. Masonry walls not directly supported on walls below require engineered design for gravity load support and design for earthquake and wind loads should be provided.

- Running board lay up of masonry units should be used instead of stack bond lay up.

Above-code Recommendations (continued):

- **For concrete masonry, use open end units at locations of vertical reinforcement and use wood beam units for horizontal reinforcing to increase the interlocking of masonry construction.**

- **Apply the measures required or recommended for masonry construction in areas of high earthquake risk, in areas of lower earthquake risk, and in high-wind areas. Priorities include provisions for reinforcing (*IRC* Figure R606.10 (2)), wall anchorage using details developed to resist out-of-plane wall loads (e.g., *IRC* Figures R611.8 (1) through (7)), minimum length of bracing walls, and a spacing limit for braced wall lines.**

Concrete and Insulating Concrete Form Wall Buildings

C NC N/A

☐ ☐ ☐ **HIGH** Construction quality control. Proper placement of reinforcing. Adequate support and attachment of reinforcing and anchor bolts. Cleaning out of cells space to allow proper concrete placement. Consolidation of concrete.

Above-code Recommendations:

- **Carefully balance bracing walls around the perimeter of the building.**

- **Apply the measures required or recommended for ICF houses in areas of high earthquake risk, in areas of lower earthquake risk, and in high-wind areas. Priorities include wall anchorage using details developed to resist out-of-place wall loads (e.g., *IRC* Figures R611.8 (2) through (7)).**

Stone and Masonry Veneer

C NC N/A

☐ ☐ ☐ **HIGH** Veneer thickness limited to 5-inch nominal thickness in SDCs A, B, and C, 4-inch nominal thickness in SDC D_1, and 3-inch actual thickness in SDC D_2 except that up to 5-inch nominal thickness can be used in SDC D_1 and D_2 if veneer only extends to first story above grade. (*IRC* Section R703.7)

Above-code Recommendations:

- Use corrosion-resistant sheet metal ties or wires to fasten veneer. The ties or wires should penetrate the house paper and sheathing and should be embedded in the wall studs.

- Where veneer can be used only on the first story above grade, increase the length of the structural wood panel bracing and use hold-down devices on braced wall panels in the first story.

Fireplaces and Chimneys

C   NC   N/A

☐   ☐   ☐   **HIGH** Masonry reinforcing. Vertical reinforcing of not less than four No. 4 bars for chimneys up to 40 x 24 inches. Should extend from bottom of foundation (3-inch minimum concrete cover) to top of chimney except that splices of not less than 24 inches are acceptable. Must be placed such that reinforcing can be surrounded in grout. Horizontal ties of 1/4-inch minimum at 18 inches maximum on center in mortar joint. SDCs D_1 and D_2. (*IRC* Section R1003.3)

Appendix C

EARTHQUAKE PROVISIONS CHECKLIST FOR DESIGNERS AND PLAN CHECKERS

General Earthquake-Resistance Requirements

General Earthquake Limitations

C   NC   N/A

☐   ☐   ☐   Seismic Design Category. Buildings in SDCs A through D_2 may be designed per the *IRC*; buildings in SDC E require engineered design unless the alternate determination of Seismic Design Category provisions of *IRC* Sections R301.2.2.1.1 or R301.2.2.1.2 are met. (*IRC* Section R301.2.2)

☐   ☐   ☐   Assembly weight. Weight of roof plus ceiling, floor, interior wall and exterior wall assemblies are limited in SDCs D_1 and D_2 and townhouses in SDC C (*IRC* Section R301.2.2.2.1).

☐   ☐   ☐   Number of stories. Wood light-frame buildings are limited to two stories plus cripple walls in SDC D_2 (*IRC* Section R301.2.2.4.1 and Table R602.10.1). Cold-formed steel framed buildings are limited to two stories above grade in SDCs D_1 and D_2 (Section R301.2.2.41). Masonry walls are limited to one story and 9 feet between lateral supports in SDCs D_1 and D_2 (*IRC* Section R606.11.3.1 and R606.11.4).

☐   ☐   ☐   Story height. All SDCs. Building story height is limited by the following limits on bearing wall clear height plus a maximum of 16 inches for the floor framing depth:

    Wood light frame   12 ft (*IRC* Section R301.3, Item 1 Exception)
    Cold-formed steel   10 ft (*IRC* Section R301.3, Item 2)
    Masonry   12 ft plus 8 ft at gable ends (*IRC* Section R301.3, Item 3)
    ICF   10 ft (*IRC* Section R301.3, Item 4, and Section 611)

Load Path

C   NC   N/A

☐   ☐   ☐   Minimum wood light frame fastening. All SDCs. (*IRC* Table R602.3)

☐   ☐   ☐   Anchor bolts and plate washers. *IRC* Section R403.1.6 for all SDCs. *IRC* Sections R403.1.6.1 and 602.11.1 for SDCs D_1 and D_2 and townhouses in SDC C.

☐   ☐   ☐   Overturning Anchorage. *IRC* Section R602.10.6 for alternate braced wall panels, all SDCs. *IRC* Section R602.10.11 Exception 2, where braced wall panels are not located at corners for SDCs D_1 and D_2. *IRC* Section R703.7, Exceptions 3 and 4, when veneer is used for SDCs D_1 and D_2.

Designed collector members aligned with and connected to the top plate of braced walls (continuous from the end of a braced wall line to the end of the braced wall panel closest to the end of the wall line). In all SDCs *IRC* Section R602.10.1. In SDC D_1 and D_2, *IRC* Section R602.10.11 last paragraph. *IRC* Section R301.2.2.2.2 for SDCs D_1 and D_2 and townhouses in SDC C.

Irregularities

C NC N/A

☐ ☐ ☐ Irregularity 1: Exterior braced wall panels not in one plane (stacked) from foundation to top most story in which they are required.

☐ ☐ ☐ Irregularity 2: Section of floor or roof not supported by braced wall lines on all edges.

☐ ☐ ☐ Irregularity 3: End of braced wall panel occurs over opening in wall below, and extends more than one foot beyond the edge of the opening.

☐ ☐ ☐ Irregularity 4: Opening in floor or roof exceeds lesser of 12 feet or 50% of least floor or roof dimension. Figure 2-x.

☐ ☐ ☐ Irregularity 5: Portions of floor level are vertically offset (split level).

☐ ☐ ☐ Irregularity 6: Braced wall lines do not occur in two perpendicular directions.

☐ ☐ ☐ Irregularity 7: Stories braced by light-frame walls include concrete or masonry construction.

Above-code Recommendations:

- **Apply irregularities to all SDCs because they are also applicable for wind load.**
- **Increase first-story strength and stiffness to mitigate weak-story irregularity.**
- **Increase cripple wall strength and stiffness to mitigate weak-story irregularity.**

Foundations and Foundation Walls

General

C NC N/A

☐ ☐ ☐ Continuous perimeter foundations. All exterior walls are to be supported on continuous perimeter foundations. All SDCs. (*IRC* Section R403.1)

☐ ☐ ☐ Continuous interior foundations. At interior braced wall lines in buildings with plan dimensions greater than 50 ft. SDCs D_1 and D_2. (*IRC* Section R403.1.2).

Appendix C, Earthquake Provisions Checklist for Designers and Plan Checkers

Special Soils Conditions

C NC N/A

☐ ☐ ☐ Low bearing capacity. Soils investigation required when building official determines that soil bearing capacities of less than 1500 psf might be present at site. All SDCs. (*IRC* Table R401.4.1, footnote b)

☐ ☐ ☐ Soil testing when expansive, compressible, or shifting soils are encountered or are likely. (*IRC* Section R401.4)

☐ ☐ ☐ Frost protection. Footings are to be below the frost line or adequate frost protection should be provided. (*IRC* Section R403.1.4.1)

Concrete Foundations

C NC N/A

☐ ☐ ☐ Minimum concrete strength. 2500 psi for all SDCs. 3000 or 3500 psi in moderate or severe weathering probability areas. (*IRC* Section R402.2 and Table R402.2)

☐ ☐ ☐ Horizontal reinforcing. One No.4 in footing and second No. 4 in stem wall. No. 4 top and bottom in thickened slab footing with alternative of one No. 5 or two No.4 in middle third of footing height for thickened slab footings cast monolithically with slab. SDCs D_1 and D_2. (*IRC* Sections R403.1.3 and R403.1.3.2)

☐ ☐ ☐ Vertical reinforcing. No. 4 at 48 inches maximum spacing required where a pour joint occurs between concrete footing and concrete stem wall. SDCs D_1 and D_2. (*IRC* Section R403.1.3)

Masonry Foundations

C NC N/A

☐ ☐ ☐ Masonry foundation type. Solid clay masonry and fully grouted concrete masonry permitted in all SDCs (*IRC* Section R403.1). Rubble stone masonry foundation walls limited to SDCs A through C (*IRC* Section R404.1.1).

☐ ☐ ☐ Horizontal reinforcing. One No. 4 in footing and second No. 4 in stem wall. SDCs D_1 and D_2. (*IRC* Section R403.1.3 and R403.1.3.1)

☐ ☐ ☐ Vertical reinforcing. Minimum No.4 at 4 feet on center extending into footing with standard hook. SDCs D_1 and D_2. (*IRC* Section R403.1.3)

Foundation Walls

C   NC   N/A

☐   ☐   ☐   Wall thickness. Six inches minimum up to 12 inches required based on soil type at site. SDCs A through D_2. (*IRC* Table R401.1.1(1)).

☐   ☐   ☐   Horizontal reinforcing. Dependent upon all thickness and material. Minimum No. 4 in upper 12 inches of wall. SDCs D_1 and D_2. (*IRC* Sections R404.1.4 and R606.11).

☐   ☐   ☐   Vertical reinforcing. Varies depending on wall height and soil type at site. ASTM Grade 60 minimum. All SDCs. (*IRC* Tables R404.1.1(2)).

Above Code Recommendations:

Avoid construction of slab-on-grade homes on cut and fill sites where possible. Where this condition cannot be avoided, provide additional quality control for fill placement and compaction operations.

Regardless of SDC, provide not less than one continuous No. 4 reinforcing bar in concrete footings. Provide a second No. 4 in stem wall if present. This will provide tension and bending capacity to help mitigate foundation damage due to earthquake, wind, soil movement, and frost heave.

In SDCs C, D_1 and D2, provide not less than No. 4 vertical bars at 4 feet as dowels between a concrete footing and separately cast slab-on-grade.

In concrete foundations, lap reinforcing bars not less than 24 inches. Bend radius (outer) for No. 4 bar is 2 inches and 2-1/2 inches for No. 5. Hook at corners and intersections of 8 inches for No.4 bars and 10 inches for No. 5 bars.

Regardless of SDC, provide not less than one continuous No. 4 reinforcing bar in masonry foundation stem walls.

In masonry foundation walls and stem walls, lap reinforcing bars not less than 24 inches. Bend radius (outer) for No. 4 bar is 2 inches and 2-1/2 inches for No. 5. Hook at corners and intersections of 8 inches for No. 4 and 10 inches for No. 5.

Floor Construction

C   NC   N/A

☐   ☐   ☐   Blocking or lateral restraint. Required at intermediate floor framing member supports. SDCs D_1 and D_2. (*IRC* Section R502.7, Exception).

Appendix C, Earthquake Provisions Checklist for Designers and Plan Checkers

Light-Frame Wall Construction

C NC N/A

☐ ☐ ☐ Braced wall length required for each 25 ft of wall length. (*IRC* Section R602.10.6)

☐ ☐ ☐ Sheathing attachment spacing. (Various *IRC* sections)

Cold-formed Steel Construction

C NC N/A

☐ ☐ ☐ Cold-formed steel framing. Buildings in SDCs D_1 and D_2 need to comply with the AISI *Standard for Cold-Formed Steel Framing – Prescriptive Method for One- and Two-Family Dwellings* in addition to the requirements of the *IRC* Section R301.2.2.4.5.

Masonry Wall Buildings

C NC N/A

☐ ☐ ☐ Limited to one story for SDCs D_1 and D_2. (*IRC* Section R606)

☐ ☐ ☐ In SDCs D_1 and D_2, 9 feet between lateral supports. (*IRC* Section R606.8)

☐ ☐ ☐ Light-frame restricted from supporting lateral loads from masonry. (*IRC* Section R301.2.2.2.2)

☐ ☐ ☐ Reinforcement detailing (*IRC* Section R606.11).

Concrete and Insulating Concrete Form Wall Buildings

C NC N/A

☐ ☐ ☐ Limited to two stories above grade. (*IRC* Section R611)

☐ ☐ ☐ Minimum wall thickness of 5.5 inches for IFC and 6 inches for solid concrete. (*IRC* Section R611.7.4)

☐ ☐ ☐ Maximum plan dimension of 60 feet and aspect ratio of 2:1. (*IRC* Section R611.2)

☐ ☐ ☐ Reinforcement detailing. (*IRC* Sections R611.3 - R611.5 and R611.7.1.2 and R611.7.1.3)

Stone and Masonry Veneer

C NC N/A
☐ ☐ ☐ Veneer. In SDCs D_1 and D_2, veneer is not permitted on buildings with cripple walls. (*IRC* Section R703.7, Exceptions 3 and 4)

Fireplaces and Chimneys

C NC N/A
☐ ☐ ☐ Vertical reinforcement requirements (four No. 4 Bars). (*IRC* Section R1003.3)
☐ ☐ ☐ Type N mortar prohibited in SDCs D_1 and D_2. (*IRC* Section R609)
☐ ☐ ☐ Anchorage requirements for SDCs D_1 and D_2. (*IRC* Section R1003.4)

Appendix D

SIGNIFICANT CHANGES FOR THE 2006 *INTERNATIONAL RESIDENTIAL CODE*

During the 2006 cycle of technical updates for the *International Residential Code* (*IRC*), a number of important technical changes were made that will have an impact on many houses around the country. This appendix highlights the most significant of these changes so that the designer can continue to use this document when the 2006 *IRC* is adopted by his or her jurisdiction.

D1 REVISED SEISMIC DESIGN MAPS

In response to concerns over the perceived increases in earthquake design forces that were implemented with the adoption of the 2000 editions of the *International Building Code* (*IBC*) and *IRC*, the U.S. Geological Survey (USGS) conducted a detailed evaluation of earthquake risk on a county-by-county basis for regions with a high probability of earthquake occurrence. The project incorporated significant new information about local geological and geotechnical features and local experts for the regions being investigated were consulted. The result of the project is revised design maps that are incorporated into the 2006 editions of the *IRC* and *IBC*. In general, the new maps reduce the amount of geographic area affected by the high seismic risk, but the remaining area is also affected so some extent. This is primarily evident in the Charleston, South Carolina, region where the Seismic Design Category was raised in the counties closest to Charleston, but the rest of the state experiences a reduction in seismic risk level. The seismic design map that was adopted for the 2006 *IRC* is shown in Figure D-1 on the following pages.

D2 ADDITION OF SEISMIC DESIGN CATEGORY D_0

In addition to provide some relief for the construction of houses using heavier finish materials such as masonry veneers and stucco, Seismic Design Category D_1 was divided into two SDCs – D_0 and D_1. Since design values must be set to the highest value in the range, dividing the original D_1 into two lowered the earthquake design load for the lower design category. Only the brick masonry veneer industry has started to take advantage of this change to date. However, it is expected that other materials also will propose changes in the future to take advantage of the lower forces associated with Seismic Design Category D_0. The geographic area associated with the change can be seen in Figure D-1.

FEMA 232, Homebuilder's Guide

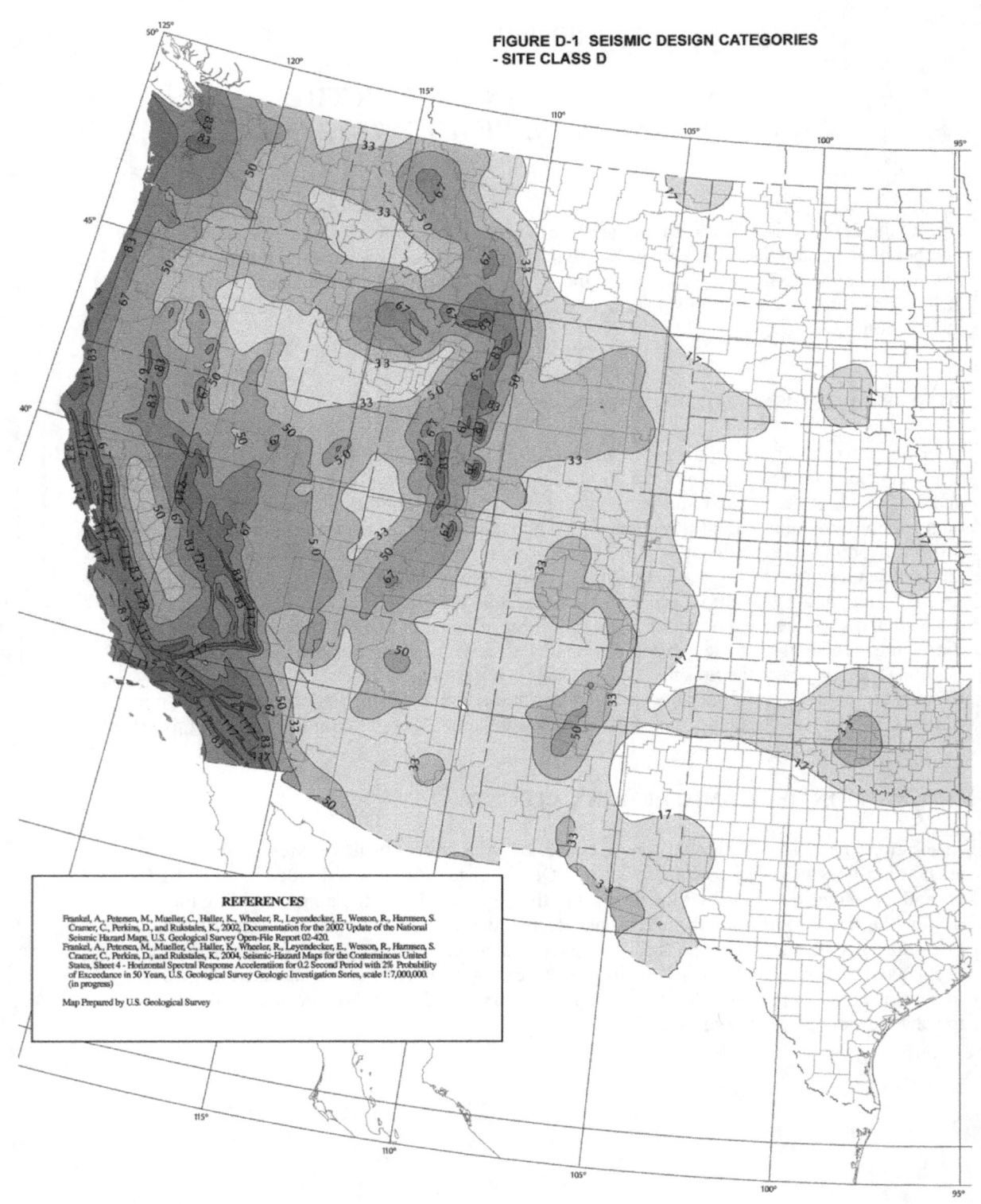

FIGURE D-1 SEISMIC DESIGN CATEGORIES - SITE CLASS D

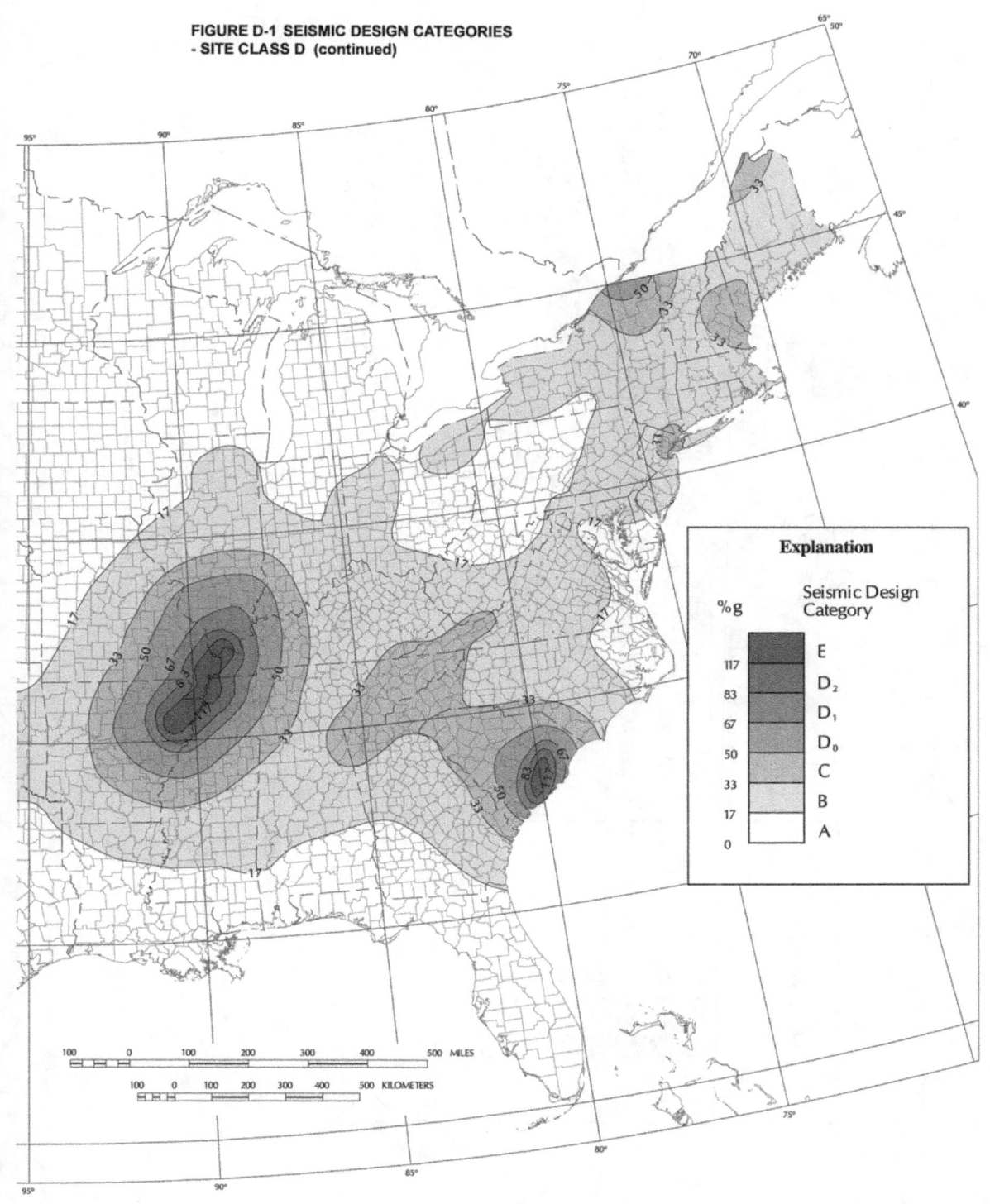

FIGURE D-1 SEISMIC DESIGN CATEGORIES - SITE CLASS D (continued)

FEMA 232, Homebuilder's Guide

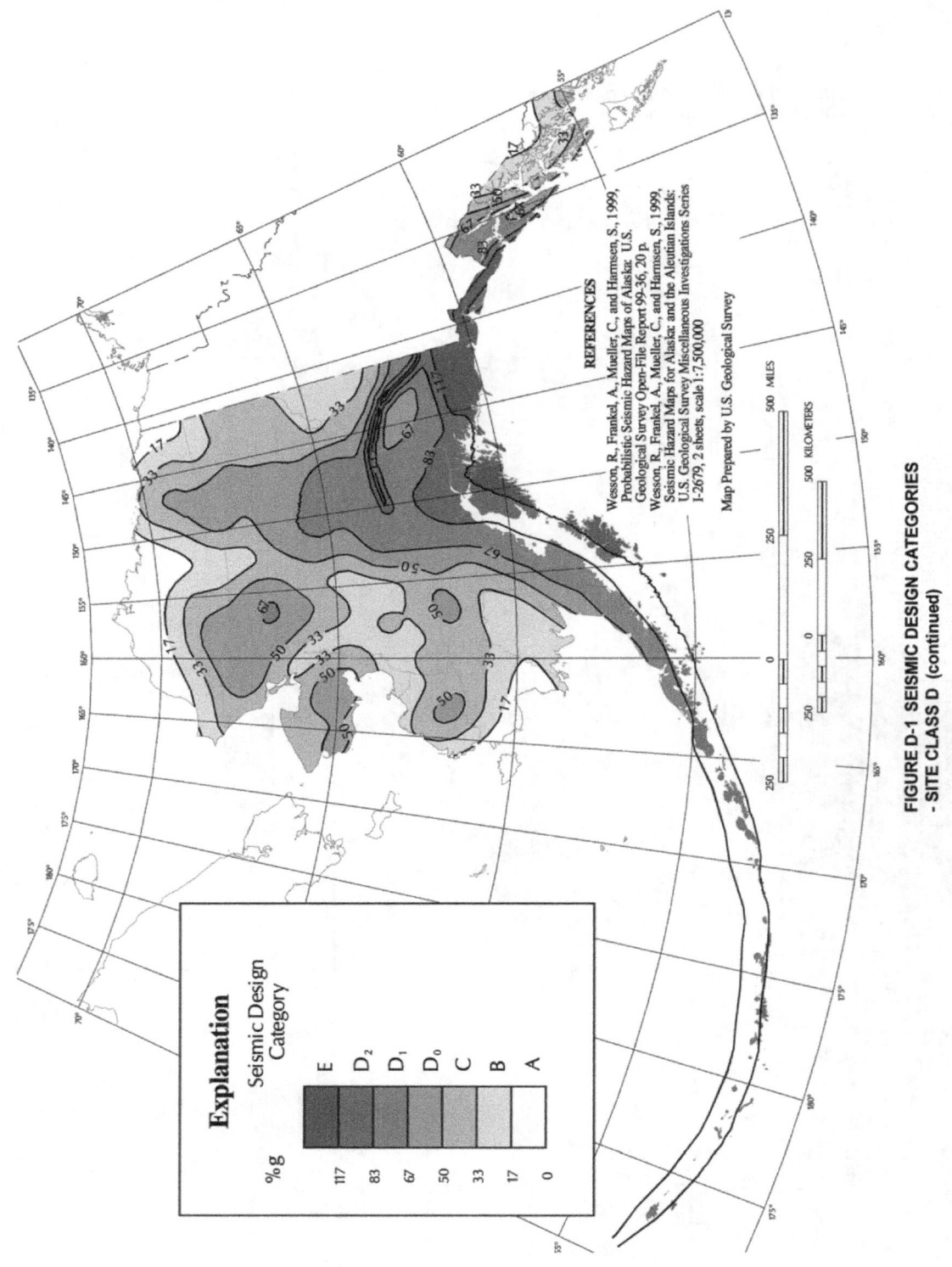

FIGURE D-1 SEISMIC DESIGN CATEGORIES - SITE CLASS D (continued)

Appendix D, Significant Changes for the 2006 *International Residential Code*

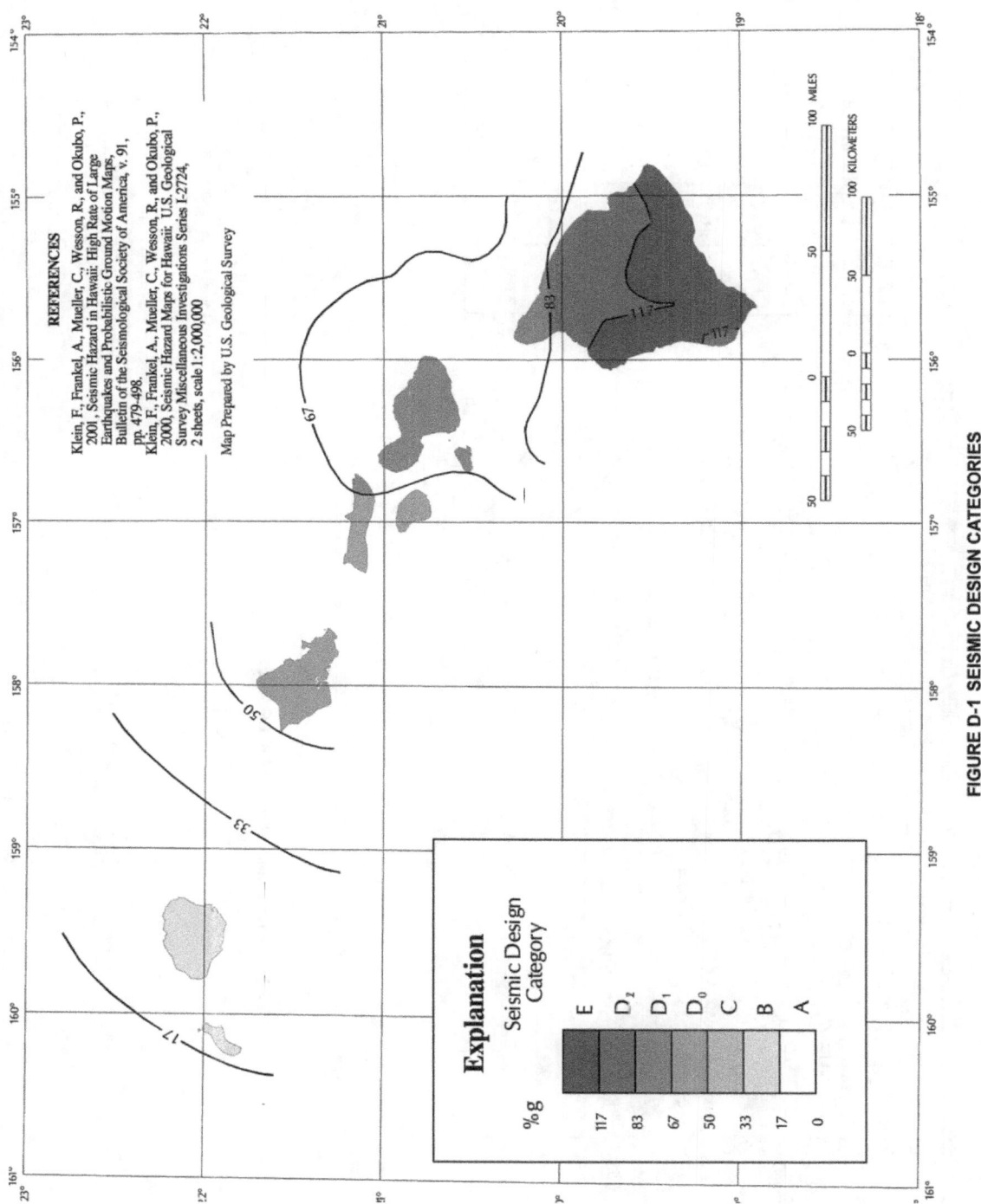

FIGURE D-1 SEISMIC DESIGN CATEGORIES - SITE CLASS D (continued)

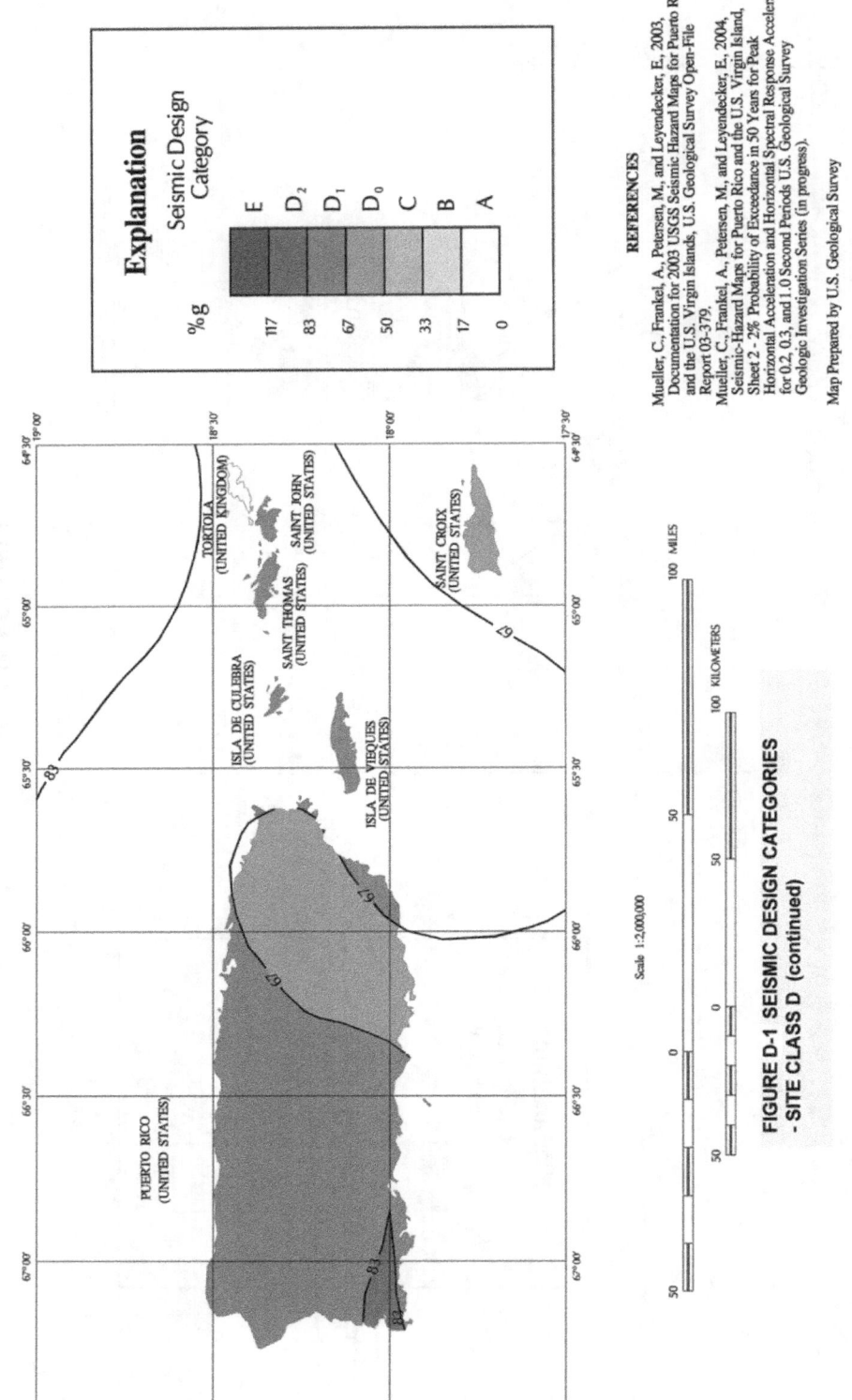

FIGURE D-1 SEISMIC DESIGN CATEGORIES - SITE CLASS D (continued)

D3 CHANGE IN APPLICABILITY OF IRREGULAR BUILDING REQUIREMENTS

During the update cycle resulting in the 2006 *IRC*, it was noticed that the wording requiring that the building conform to the requirements of *IRC* Section R301.2.2.2.2 only applied to wood and ICF concrete construction. Cold-formed steel construction had to conform to these irregularity requirements even through the referenced design and construction document (COFS/PM) had less stringent requirements and application to masonry wall buildings was not specified. Thus, a change was made to require all buildings to conform to the irregularity provisions of the *IRC*, which limit the concentrations of loads and deformations that irregularities cause.

D4 CLARIFICATION AND ADDITION OF REQUIREMENTS FOR MASONRY VENEER

A significant change was made to the requirements for the use of stone and masonry veneer in areas of high earthquake risk. The changes clarify veneer weight limits and stories where veneer is permitted. They also clarify and illustrate the required hold-down anchorage of walls and the requirements for ties and other reinforcement and attachment of the veneer to the walls. The change is too extensive to document here; rather, the reader is referred to the 2006 *IRC* for details.

Appendix E
REFERENCES AND ADDITIONAL RESOURCES

American Forest & Paper Association (AF&PA). 2001. *Wood Frame Construction Manual for One- and Two-Family Dwellings* and *Commentary*. AF&PA, Washington, D.C.

American Forest & Paper Association. 2001. *Allowable Stress Design (ASD) Manual for Engineered Wood Construction*. AF&PA, Washington, D.C.

American Forest & Paper Association. 1996. *Load and Resistance Factor Design (LRFD) Manual for Engineered Wood Construction*. AF&PA, Washington, D.C.

American Forest & Paper Association. 2001. *Details for Conventional Wood frame Construction*. AF&PA, Washington, D.C.

American Forest & Paper Association. 1993. *Span Tables for Joists and Rafters*. AF&PA, Washington, D.C.

American Forest & Paper Association. 1987. *Basic Requirements for Permanent Wood Foundation System*, Technical Report 7. AF&PA, Washington, D.C.

American Institute of Timber Construction (AITC). 2004. *Timber Construction Manual – Fifth Edition*. AITC, Englewood, Colorado

American Iron and Steel Institute (AISI). 2001. *Standard for Cold-Formed Steel Framing - Prescriptive Method for One- and Two-Family Dwellings (AISI/COFS/PM 2001)*. AISI, Washington, D.C.

Anderson, C., F. Woeste, and J. Loferski. 2005. *Manual for the Inspection of Residential Wood Decks and Balconies*. ICC, Country Club Hills, Illinois, and Forest Products Society, Madison, Wisconsin.

Applied Technology Council (ATC). 1976. *A Methodology for Seismic Design and Construction of Single-Family Dwellings,* ATC 4. ATC, Redwood City, California.

Association of Bay Area Governments (ABAG). *Training Materials for Seismic Retrofit of Wood-Frame Homes*, ABAG, Oakland, California.
www.abag.ca.gov/bayarea/eqmaps/fixit/training.html

Association of Bay Area Governments (ABAG). *Info on Chimney Safety and Earthquakes*. ABAG, Oakland, California. www.abag.ca.gov/bayarea/eqmaps/fixit/chimneys.html.

Association of Bay Area Governments. *Training Materials for Seismic Retrofit of Wood-Frame Homes*. Available online at http://www.abag.ca.gov/bayarea/eqmaps/fixit/training.html.

Breyer, D. E., K. J. Fridley, D. G. Pollock, K. E. Cobeen. 2003. *Design of Wood Structures – ASD*. McGraw-Hill Publishing, New York, New York.

Building Seismic Safety Council (BSSC). Website: http://www.bssconline.org. This site provides links to the BSSC's 60+ member organizations' websites.

California Seismic Safety Commission. 1992. *The Homeowner's Guide to Earthquake Safety*. Seismic Safety Commission, Sacramento, California.

California Division of the State Architect (DSA). 2002. *Guidelines for Earthquake Bracing of Residential Water Heaters*. DSA, Sacramento, California.

Canadian Wood Council (CWC). 1991. *Wood Reference Handbook*. CWC, Ottawa, Ontario, Canada.

Canadian Wood Council. 1993. *Wood Building Technology*. CWC, Ottawa, Ontario, Canada.

Chen W.-F. and C. Scawthorn. 2004. *Earthquake Engineering Handbook*, Second Edition. CRC Press, LLC, Boca Raton, Florida.

City of Los Angeles, California. 2001. *Reconstruction and Replacement of Earthquake Damaged Masonry Chimneys,* Information Bulletin P/BC 2001-70.

City of Los Angeles, California. 2002. *City of Los Angeles Building Code*, Chapter 94 "Voluntary Earthquake Hazard Reduction in Existing Hillside Buildings."

City of San Leandro, California. *Homeowner's Handbook and Home Strengthening Plan Set*. www.ci.san-leandro.ca.us/homehandbook.html.

City of Seattle, Washington. *Project Impact Home Retrofit Series*. www.ci.seattle.wa.us/projectimpact/pages/publications/homeretrofitseries.htm.

Consortium of Universities for Research in Earthquake Engineering. *CUREE-Caltech Woodframe Project*. A number of reports are available; consult the project website: https://secure.curee.org/catalog/index.php?main_page=index&cPath=3.

Earthquake Engineering Research Institute (EERI)/International Conference of Building Officials (ICBO). *Resisting the Forces of Earthquakes* (video).

Faherty, K. F., and T. G. Williamson. 1999. *Wood Engineering and Construction Handbook – Third Edition*. McGraw-Hill Publishing, New York, New York

Federal Emergency Management Agency (FEMA). Website: http://www.fema.gov.

Federal Emergency Management Agency. *Above the Flood: Elevating Your Flood-prone House*, FEMA 347. FEMA, Washington, D.C.

Federal Emergency Management Agency. *Coastal Construction Manual: Principles and Practices of Planning, Siting, Designing, Constructing, and Maintaining Residential Buildings in Coastal Areas*, FEMA 55. FEMA, Washington, D.C.

Federal Emergency Management Agency. *Design Guidelines for Flood Damage Reduction*, FEMA 15. FEMA, Washington, D.C.

Federal Emergency Management Agency. *Elevated Residential Structures*, FEMA 54. FEMA, Washington, D.C.

Federal Emergency Management Agency. *FEMA Home Builders Guide to Coastal Construction: Technical Fact Sheet Series*, FEMA 499. FEMA, Washington, D.C.

Federal Emergency Management Agency. *Homeowners Guide to Retrofitting: Six Ways to Protect Your House from Flooding*, FEMA 312. FEMA, Washington, D.C.

Federal Emergency Management Agency. *Reducing Flood Losses Through the International Codes: Meeting the Requirements of the National Flood Insurance Program*. FEMA, Washington, D.C., and ICC, Country Club Hills, Illinois.

Federal Emergency Management Agency. *Reducing the Risks of Nonstructural Earthquake Damage*, FEMA 74. FEMA, Washington, D.C.

Federal Emergency Management Agency. *Taking Shelter from the Storm: Building a Safe Room Inside Your House*, FEMA 320. FEMA, Washington, D.C.

Folz, B., and A. Filiatrault. 2002. *A Computer Program for Seismic Analysis of Woodframe Structures*, CUREE W-21. Consortium of Universities for Research in Earthquake Engineering, Richmond, California.

Gurfinkel, G.. 1973. *Wood Engineering*. Southern Forest Products Association, New Orleans, Louisiana.

International Code Council (ICC). Website: http://www.iccsafe.org.

International Code Council (ICC). 2003a. *International Residential Code*. ICC, Country Club Hills, Illinois.

International Code Council (ICC). 2003b. *International Building Code*. ICC, Country Club Hills, Illinois.

International Code Council (ICC). 2003c. *International Existing Building Code* (IEBC). ICC, Country Club Hills, Illinois.

International Code Council (ICC). 2003d. *Conventional Construction Provisions of the 2003 IBC, An Illustrated Guide.* ICC, Country Club Hills, Illinois.

Masonry Institute of America (MIA). 1995. *Masonry Fireplace and Chimney Handbook.* MIA, Torrance, California.

International Code Council (ICC). 2005. *International Existing Building Code Commentary.* ICC, Country Club Hills, Illinois.

Masonry Standards Joint Committee (MSJC). 2002. *Masonry Standards Joint Committee Code, Specification, and Commentaries.* ACI International, Structural Engineering Institute, The Masonry Society, Washington, D.C.

McClure, F. E. 1973. *Performance of Single Family Dwellings in the San Fernando Earthquake of February 9, 1971.* U.S. Department of Commerce, National Oceanic and Atmospheric Administration, Washington, D.C.

National Fire Protection Association (NFPA). 2006. *Building Construction and Safety Code*, NFPA 5000. NFPA, Quincy, Massachusetts.

Porter, Keith A., Charles R. Scawthorn, and James L. Beck. 2006. "Cost-effectiveness of Stronger Woodframe Buildings." *Earthquake Spectra*, 22(February):1, 0pp 239-266.

Slosson, J. E. 1975. "Effects of the Earthquake on Residential Areas." Chapter 19 of *San Fernando, California, Earthquake of 9 February*, California Division of Mines and Geology, Bulletin 196.

Southern Building Code Congress International. 1999. *Standard for Huricane-Resistant Residential Construction*, SSTD 10. International Code Council, Country Club Hills, Illinois.

Stewart, J. P., J. D. Bray, R. B. Seed, and N. Sitar, Eds. 1994. *Preliminary Report on the Principal Geotechnical Aspects of the January 17, 1994, Northridge Earthquake*, UCB/EERC-94/08. Earthquake Engineering Research Center, University of California, Berkeley.

Stewart, J. P., J. D. Bray, D. J. McMahon, and A. L. Kroop. 1995. "Seismic Performance of Hillside Fills." In *Landslides under Static and Dynamic Conditions – Analysis Monitoring and Mitigation*, Geotechnical Special Publication 52, edited by C. L. Ho and D. K. Keefer. American Society of Civil Engineers, Reston, Virginia.

Structural Engineers Association of California. 2003. *Commentary on Chapter 3 Guidelines for Seismic Retrofit of Existing Buildings, 1 October 2003 Version.* Existing Buildings Committee, SEAOC, Sacramento, California.

Structural Engineers Association of Northern California (SEAONC). 2001. *Guidelines for Seismic Evaluation and Rehabilitation of Tilt-up Buildings and Other Rigid Wall/Flexible Diaphragm Structures.* SEAONC, San Francisco, California.

U.S. Department of Agriculture (USDA). 1999. *Wood Handbook – Wood as an Engineering Material*, USDA Forest Products Laboratory General Technical Report FPL-GTR-113. USDA Forest Products Laboratory, Madison, Wisconsin.

Von Winterfeldt, D., N. Roselund, and A. Kitsuse. 2000. *Framing Earthquake Retrofitting Decisions: The Case of Hillside Homes in Los Angeles*, PEER Report 2000-03. Pacific Engineering Research Center, Richmond, California.

Wood Truss Council of America (WTCA). 1997. *Metal Plate Connected Wood Truss Handbook – Third Edition.* WTCA, Madison, Wisconsin.

Appendix F
HOMEBUILDERS' GUIDE PROJECT PARTICIPANTS

BSSC HOMEBUILDERS' GUIDE PROJECT COMMITTEE

Chair – J. Daniel Dolan, PhD, PE, Professor, Washington State University, Wood Materials and Engineering Laboratory, Pullman (writing team leader)

Members
Kelly Cobeen, Structural Engineer, Cobeen and Associates Structural Engineering, Lafayette, California (co-author)
Gerald Jones, PE, Code Official, Retired, Kansas City, Missouri
James E. Russell, Building Codes Consultant, Concord, California (co-author)
Jim W. Sealy, FAIA, Architect/Consultant, Dallas, Texas
Douglas M. Smits, CBO, City of Charleston, South Carolina

FEMA Liaison Members
Robert D. Hanson, Federal Emergency Management Agency, Walnut Creek, California
Michael G. Mahoney, Federal Emergency Management Agency, Washington, D.C.

INVITEES/PARTICIPANTS IN HOMEBUILDERS' GUIDE WORKSHOPS AND REPORT REVIEWERS

Mamood Abolhoda, City of Freemont, California
Thomas Ahrens, City of Mill Valley, California
John C. Anderson, City of San Diego, California
Paul Armstrong, International Code Council, Whittier, California
Christopher Arnold, FAIA, RIBA, Building Systems Development Inc., Palo Alto, California
James Bela, Oregon Earthquake Awareness
Carol Bloom, State Farm Insurance, Rohnert Park, California
J. Gregg Borchelt, PE, Brick Industry Association, Reston, Virginia
Amanda Brady, City of San Diego, California
Patrick Bridges, Builder, Portland, Oregon
Mark Caldwell, Apex Engineers, Calvert City, Kentucky
Lee Campbell, Property Solutions, Inc., Irvine, California
Cathleen Carlisle, Federal Emergency Management Agency, Washington, D.C.
Tom Caulfield, Muller and Caulfield Architects, Oakland, California
Frank Chiu, City and County of San Francisco, California
Michael Christoffersen, Architectural Designs, Inc., Fort Wayne, Indiana
John Chrysler, Masonry Institute of America, Torrance, California
Glen Clapper, Southern Brick Institute, Conyers, Georgia
Nicolino G. Delli Quadri, SE, City of Los Angeles, California
Bradford K. Douglas, PE, American Forest and Paper Association, Washington, D.C.

Ali Fattah, City of San Diego, California
Jeffrey K. Feid, State Farm Insurance Company, Bloomington, Illinois
Art Garcia, City of San Diego, California
Robert S. George, Architect, South San Francisco, California
Frank Golon, Pulte Homes, Bernardsville, New Jersey
Rose Grant, State Farm Insurance, Bloomington, Illinois
Michael Gregor, SGM Architecture Interiors, Mount Pleasant, South Carolina
Cynthia Hammond, CPCU, State Farm Insurance, Rohnert Park, California
Isam Hasenn, City of San Diego, California
Perry A. Haviland, FAIA, Haviland Associates Architects, Oakland, California
John R. Henry, PE, International Code Council, Sacramento, California
Fred M. Herman, City of Palo Alto, California
George Horton, Oakland, California
Eric N. Johnson, Brick Association of the Carolinas, Charlotte, North Carolina
Stephan A. Kiefer, City of Livermore, California
Joseph Knarich, National Association of Home Builders, Washington, D.C.
Vladimir G. Kochkin, National Association of Home Builders Research, Upper Marlboro, Maryland
Laurence Kornfield, City and County of San Francisco, California
Ray Kothe, Builder, Baton Rouge, Louisiana
Jay W. Larson, American Iron and Steel Institute, Bethlehem, Pennsylvania
Sheila Lee, City of Santa Clara, California
Adlai Leiby, Berkeley, California
Philip Line, American Forest and Paper Association, Washington, D.C.
Jeff Lusk, Federal Emergency Management Agency Region IX, Oakland, California
Joan O. MacQuarry, City of Berkeley, California
Bonnie Manley, National Fire Protection Association, Quincy, Massachusetts
Harry W. Martin, American Iron and Steel Institute, Auburn, California
Zeno A. Martin, PE, APA-The Engineered Wood Association, Tacoma, Washington
Joseph J. Messersmith, Jr., PE, Portland Cement Association, Rockville, Virginia
Greg Mulvey, City of San Diego, California
Steve Pfeiffer, City of Seattle, Seattle, Washington
David Pollock, Washington State University, Pullman
Steven E. Pryor, Simpson Strong-Tie, Dublin, California
Kevin Reinerston, HCD Division of Codes and Standards, Sacramento, California
Timothy A. Reinhold, PhD, PE, Institute for Business and Home Safety, Tampa, Florida
Dennis Richardson, San Jose, California
Alan Robinson, Tuan and Robinson Structural Engineers, Inc., San Francisco, California
William R. Schock, City of San Leandro, California
Roger Sharpe, City of Walnut Creek, Retired, Berkeley, California
Thomas D. Skaggs, PhD, PE, APA-The Engineered Wood Association, Tacoma, Washington
Steve Skalko, Portland Cement Association, Macon, Georgia
Charles A. Spitz, AIA, CSI, NCARB, Architect-Planner-Code Consultant, Wall, New Jersey
William W. Stewart, FAIA, Stewart-Schaberg/Architects, Chesterfield, Missouri
Jason Thompson, SE, National Concrete Masonry Association, Herndon, Virginia
Ray Tu, Hardy Frames, Inc., Ventura, California

Fred M. Turner, Seismic Safety Commission, Sacramento, California
David P. Tyree, PE, CBO, American Forest and Paper Association, Colorado Springs, Colorado
Calvin N. Wong, PE, City of Oakland, California
Matt Zamani, City of San Diego, California

2005-2006 BSSC BOARD OF DIRECTION

Chair -- Jim W. Sealy, FAIA, Architect/Consultant, Dallas, TX

Vice Chair -- David Bonneville, Degenkolb Engineers, San Francisco, California

Secretary -- Jim Rinner, Project Manager II, Kitchell CEM, Sacramento, California

Ex-Officio Member -- Charles Thornton, Chairman/Principal, Thornton-Tomasetti Group, Inc., New York, New York

Members
Edwin Dean, Nishkian Dean, Portland, Oregon
Bradford K. Douglas, Director of Engineering, American Forest and Paper Association, Washington, D.C.
Cynthia J. Duncan, Director of Specifications, American Institute of Steel Construction, Chicago, Illinois
Henry Green, Executive Director, Bureau of Construction Codes and Fire Safety, State of Michigan, Department of Labor and Economic Growth, Lansing, Michigan (representing the National Institute of Building Sciences)
Jay W. Larson, American Iron and Steel Institute, Bethlehem, Pennsylvania
Joseph Messersmith, Coordinating Manager, Regional Code Services, Portland Cement Association, Rockville, Virginia (representing the Portland Cement Association)
Ronald E. Piester, Assistant Director for Code Development, New York State, Department of State, Kinderhook, New York
James Rossberg, Manager, Technical Activities for the Structural Engineering Institute, American Society of Civil Engineers, Reston Virginia
W. Lee Shoemaker, Director, Engineering and Research, Metal Building Manufacturers Association, Cleveland, Ohio
Howard Simpson, Simpson Gumpertz and Heger, Arlington, Massachusetts (representing National Council of Structural Engineers Associations)
Shyam Sunder, Deputy Director, Building Fire Research Laboratory, National Institute of Standards and Technology, Gaithersburg, Maryland (representing Interagency Committee on Seismic Safety in Construction)
Charles A. Spitz, Architect/Planner/Code Consultant, Wall New Jersey (representing the American Institute of Architects)
Robert D. Thomas, Vice President Engineering, National Concrete Masonry Association, Herndon, Virginia

BSSC Staff
Claret M. Heider, Vice President for BSSC Programs
Bernard F. Murphy, PE, Director, Special Projects
Carita Tanner, Communications/Public Relations Manager

Appendix F, Homebuilders' Project Participants

THE BUILDING SEISMIC SAFETY COUNCIL

The purpose of the Building Seismic Safety Council is to enhance the public's safety by providing a national forum to foster improved seismic safety provisions for use by the building community. For the purposes of the Council, the building community is taken to include all those involved in the planning, design, construction, regulation, and utilization of buildings.

To achieve its purposes, the Council shall conduct activities and provide the leadership needed to:

- Promote development of seismic safety provisions suitable for use throughout the United States;

- Recommend, encourage, and promote adoption of appropriate seismic safety provisions in voluntary standards and model codes;

- Assess implementation progress by federal, state, and local regulatory and construction agencies;

- Identify opportunities for the improvement of seismic regulations and practices and encourage public and private organizations to effect such improvements;

- Promote the development of training and educational courses and materials for use by design professionals, builders, building regulatory officials, elected officials, industry representatives, other members of the building community and the public.

- Provide advice to governmental bodies on their programs of research, development, and implementation; and

- Periodically review and evaluate research findings, practice, and experience and make recommendations for incorporation into seismic design practices.

The scope of the Council's activities encompasses seismic safety of structures with explicit consideration and assessment of the social, technical, administrative, political, legal, and economic implications of its deliberations and recommendations.

Achievement of the Council's purpose is important to all in the public and private sectors. Council activities will provide an opportunity for participation by those at interest, including local, State, and Federal Government, voluntary organizations, business, industry, the design professions, the construction industry, the research community and the public. Regional and local differences in the nature and magnitude of potentially hazardous earthquake events require a flexible approach adaptable to the relative risk, resources and capabilities of each community. The Council recognizes that appropriate earthquake hazard reduction measures and initiatives should be adopted by existing organizations and institutions and incorporated into their legislation, regulations, practices, rules, codes, relief procedures and loan requirements, whenever possible, so that these measures and initiatives become part of established activities rather than being superposed as separate and additional.

The Council is established as a voluntary advisory, facilitative council of the National Institute of Building Sciences, a nonprofit corporation incorporated in the District of Columbia, under the authority given the Institute by the Housing and Community Development Act of 1974, (Public Law 93-383), Title VIII, in furtherance of the objectives of the Earthquake Hazards Reduction Act of 1977 (Public Law 95-124) and in support of the President's National Earthquake Hazards Reduction Program, June 22, 1978.

www.ingramcontent.com/pod-product-compliance
Lightning Source LLC
Chambersburg PA
CBHW082118230426
43671CB00015B/2734